Keys to Excellence

Seventh Edition

Edited by
Carol Cooper, Ed.D.

KENDALL/HUNT PUBLISHING COMPANY
4050 Westmark Drive   Dubuque, Iowa 52004

Photographer. Joseph Ricco, photo page—Dario Mercado
     (some photos provided by authors)
Illustrator. Mary Nichols, Illustrations—chapter 9, Dr. Len Koeth

Cover image courtesy of Stockbyte ©2004.

Contents

Preface

To the Student

Hello and welcome to the beginning of your college education. Congratulations on having decided to begin it here. Consider these years as the period that is probably going to shape the rest of your life. It is a time when you will grow emotionally and intellectually. You will meet other people, learn about interacting with other cultures, perhaps question your values, and further sharpen your critical thinking and academic skills. It is an exciting time. Enjoy it.

To prepare you for this experience, we have developed this orientation text to help you become familiar with and adjusted to college life. We want to help you meet all of these new challenges as you mature. We want you to be a successful student and get the best education possible.

At the college we endeavor to provide for your needs by making all of our resources available to you. We even include career assessments that if taken outside the college would cost hundreds of dollars. How do we know what to make available to you? Well, over the years, we have talked with many students and graduates about the orientation course. They have continually provided us with feedback about what should and should not be included in the course's curriculum. We have also done research studies comparing students who took the orientation course with those who did not. Guess what, each time the results show students who take the course have a higher graduation rate than those who do not and/or were not successful in the course.

The freshman year of college is crucial. The greatest percentage of students who drop out do so during their first year of school. Therefore, this course is designed to get you through your first college year. With this course, your odds for finishing school will not only be greater, but now you will have the academic skills to get you through the rest of your college years.

The information in this text will help you develop life-and college-coping skills. **It is a guide**. It will not work magic or bring you luck. You have to be willing to put in the necessary effort to make yourself successful. As you go through the text and course, we welcome your comments, both positive and negative, to help keep us on track. Feel free to share this information with your instructor or use the e-mail address of the authors to share your feedback. Remember, you can be successful, but only if you really put in the best effort for you. You can be successful if you follow the *Keys to Excellence* presented in this text.

TEXT SUMMARY

This text is designed for you to actively participate by completing exercises in each of the twelve chapters. We have included numerous exercises because we have found that when students actively participate, they learn more skills and become more adept at using them. The chapters include:

Tips on Adjusting to College

Campus Resources

Library & Internet Information

Motivation & Success

Time Management

Listening & Note taking

Memory

The Art of Test Taking

Career Planning for the 21st Century

Academic Regulation

Diversity in Higher Education

Wellness

In these chapters we start off with an awareness check to see where you think you are in terms of knowledge on a particular skill and end with a summary quiz. In some instances, you will be asked to share your personal experiences and work in small groups. Before this course is over, if you so choose, you will have further developed career goals, selected and researched careers and know exactly what courses you will need to take in order to complete your college education. We also make available course information for students transferring to upper division schools— universities. Wellness is listed last, but is not considered as the area in which you need the least amount of knowledge. You need to be in good physical and mental shape to cope effectively with the eleven other keys and college in general.

As an addition to this text, we have designed it so you can keep up with life changes you make or need to make through a **Journal** that the instructor will read and provide you with feedback to further assist you. Each chapter will provide you with questions, activities and/or Internet links for you to response to in your journal. Buy a three-hole punch folder with pockets that will hold 8½ x 11 paper. Each journal entry page should be headed with the chapter title, and date.

In the appendix, we have included a student information sheet that you will need to complete during the first week of class and return to your instructor. This is one of the ways that he or she will use to keep in touch with you. We have also provided forms to help you map out the courses you need and when you will take them.

Your instructor will act as your mentor during your first term here. They will also be available to assist you with registration for the next term.

There are many skills you need to develop as a student and you need them now. However, since it is impossible to share all of them at once, in chapter one, we have provided you with information we believe will be helpful to you until you can learn those skills. Some of the information is based on common sense, but it is mentioned to bring it to the forefront of your consciousness.

To the Instructor

This is the seventh edition and it comes with a student self-assessment, and learning style inventory along with many website supports for you to encourage your students to use. In chapter three, we have provided "how to" information on the Internet along with the addresses of several search engines such as altavista.com, lycos.com, yahoo.com and many others. The exercises provided in each chapter are ideal for classroom use as well as homework. We have asked the students to keep a journal of their life changes as they go through their first term. We have also suggested that if they have issues they want you as the instructor to comment on, that they should also write them in their journals. Journal questions and activities have been provided to guide them. Feel free to ask students to include other information. You may want to use the journal as a writing project or to simply collect it as a midterm and final project.

For classroom strategies or support, contact one of the authors listed below based on the chapter.

Carol Cooper, Ed.D.   ccooper @mdcc.edu
Sheryl Hartman, Ph.D.   shartman@mdcc.edu
Sandra Schultz, Ph.D.   sschultz@mdcc.edu

Acknowledgments

I wish to express my sincere gratitude to all of the staff and faculty of Miami-Dade Community College for their contributions. My special thanks go to the current and former students whose feedback resulted in many of the improvements which are included in the text.

I also wish to thank the following reviewers: Yvonnie Coffie, Advisement Services; Wiley Huff, Professor, Criminal Justice; Shelley E. Davis, Professor, Social Sciences; Marina Gonzalez, Academic Affairs.

Name _____   Date _____

Assessment Skills

Directions: Mark your answer on the chart at the appropriate level (based on your score) for each key

Example: See the **example** below. This person scored 13 on key 1 (Time Management) and 11 on key 2 (Motivation Success)

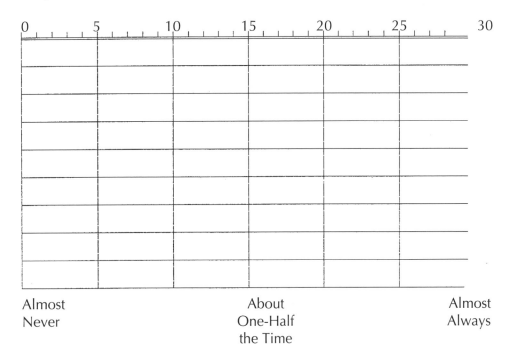

Frequency of Behavior
Example

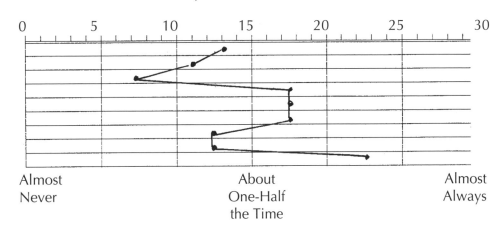

Name _____   Date _____

DIRECTIONS:

Use this page to (1) list things you may have heard about college but need to have clarified, and (2) list other topics of interest you believe should be covered in this college success class.

College   **Other Topics of Interest**

1. _____   2. _____

_____   _____

_____   _____

_____   _____

_____   _____

_____   _____

_____   _____

_____   _____

_____   _____

_____   _____

_____   _____

_____   _____

_____   _____

_____   _____

Chapter One

Academic Excellence

Carol Cooper, Ed.D.

The way to excel

Exercise 1.1

Academic Excellence Awareness Check

DIRECTIONS: Please place an "X" in the appropriate box.

	Yes	No	
1.	☐	☐	College is a teaching environment.
2.	☐	☐	A syllabus is your course guide.
3.	☐	☐	Your instructor is responsible for what you learn in the course.
4.	☐	☐	Sitting in front of the classroom can help you earn a higher grade.
5.	☐	☐	Your instructor's expectations are listed in the course syllabus.
6.	☐	☐	Test preparation begins on the first day of class.
7.	☐	☐	In college, students do not really have to come to class.
8.	☐	☐	A student must sacrifice in order to achieve.
9.	☐	☐	Always skim through your books before class.
10.	☐	☐	Joining a study group or club makes you more likely to succeed in school.

Introduction

You are now a freshman in your first semester at college. You have no doubt taken a placement test and been advised into courses based on your abilities. Now you are pondering if you are going to get what you need out of college to succeed in the world of work. First, you need to understand what college is all about and what it seeks to do.

Colleges are institutions of higher learning where faculty are given academic freedom to educate students. Faculty teach and/or provide formal instruction, advise, act as role models, study, do research, motivate and try to enhance your mental and psychological growth. Their major tasks are to assist the college in producing educated people usually in vocational and/or white collar professions and to help students get to where they want to go. Faculty strive to help students learn at a higher level than they did in high school. They are genuinely interested in helping you achieve your career goals.

At the community college level, degrees are offered at the associate level. The most common are:

> Associate in Arts (A.A.)
> Associate in Science (A.S.)

Community colleges also offer training in vocational areas that will provide you with certificates of completion. They are referred to as: advanced technical, college credit and vocational credit programs.

Once you move into higher education from the community college level into, usually, four-year colleges, and universities, the following are some of the most common degrees awarded:

> Bachelor's—undergraduate
>> arts (B.A.)
>> sciences (B.S.)
>
> Master's—graduate
>> arts (M.A.)
>> sciences (M.S.)
>
> Doctorates—graduate
>> philosophy (Ph.D.)
>> education (Ed.D.)

To earn these degrees in a timely manner, you have to organize your plan of study at least by the end of the first term in school. Early career planning is the key to effective academic course scheduling. See general education requirements in your college catalog or it can be found on the Internet at http://www.mdc.edu, then click on Student Information Services if you are attending school at Miami Dade College. When you are focused, it is just a matter of achieving your career goals through hard work. College should be considered just like a full-time job. It is a big business. If you invest your time now, it will provide huge dividends in the

Welcome to higher education

College faculty

Career Planning

future. Choose a major that will put you on the right road to your career. Don't leave your education and career to chance. Give it your best shot.

College and income information

More than one-half of students enrolling in college start at two-year institutions—so you are in good company. According to the American Association of Community Colleges, the average age of community college students is 29-years-old. More than half of those older than 25 are employed full time. Approximately one-half of starting students will drop out. Don't be in the half that drops out. There is good news. If you are in the half that stay, your "average annual earnings with an associate degree is almost $7000 more than a person with just a high school diploma." This means that over a lifetime you can earn more than $250,000 over the high school graduate.

Characteristics of an educated person

The educated person is considered to have some of the following characteristics:

- Open-minded

- Well-rounded

- Honest

- Ability to adapt to change

- Conversant on where they are

- Critical thinker

- Problem-solver

- Self-reliant

- Studied in a particular field

- Responsible

- Self-supporting

- Life-long learner

- Sense of social obligation to assist in the betterment of society

- Continually striving to reach higher levels of understanding

Your instructor certainly cannot make you into an educated person in this orientation course, but he/she can help you begin the process and assist you in setting goals which will guide you throughout your educational experiences. Your instructor will introduce you to some of the techniques you will need to master if you want to be successful in college and life in general. This text will also assist you in dealing with obstacles you may encounter along the way and provide you with strategies on how to resolve them.

Skills needed for college

The authors of this text have taken the time to review research on what skills students need to be successful. The vast majority of the researchers have concluded that if students learned how to organize and manage their time; understood the role of motivation, goal-setting and

prioritizing; and used effective study skills such as note-taking and test taking, they would be successful. Another important skill is "being sensitive to other cultures in the environment, yet being effective in their learning approach." As a student, you cannot allow instructor/peer differences to hinder you from achieving academic success.

What are some of your responsibilities in college?

Student responsibility

- Come to class prepared

- Learn and use critical thinking

- Follow directions

- Read & Listen

- Know your learning style and how to use it effectively

- Learn your strengths & weaknesses

- Ask for help when needed

Before you finish this chapter, you will be asked to stop and take a Learning Style Inventory (LSI). It will provide you with information on your preferred learning style and what learning strategies you should use in the classroom if you want to be successful.

Research also showed that most college students needed help in making the transition from high school to college.

Teaching to Learning

There is one key you need to understand to get you started—the difference between a *high school environment* and a *college environment*. Understanding and following through on the change in environments can make the difference as to whether you will be here or not at the end of the school year.

According to Siebert and Walter, high school is a *teaching environment* and college is a *learning environment*. In high school, "how much you learned was strongly determined by the skills of your teachers."[1] Not only that, they constantly reminded you when to read, write and hand in assignments. In college, "the responsibility for what you learn is yours, not the instructor's."[2] In most classes at the beginning of the term, instructors will provide you with a *syllabus* that outlines your course responsibilities. You may have a few reminders, but basically, it is now up to you. You must ask the questions if you don't understand and get clarity on assignments. You must make it your responsibility to understand your instructors and their expectations of you. Assume that instructors are responsible for

Teaching environment/ Learning environment

Syllabus

---

1. Tim Walter and Al Siebert. *Student Success: How to Succeed in College and Still Have Time for Your Friends*, 5th ed., Harcourt Brace Jovanovich, Inc. 1990, p. 4

2. *Ibid.*, p. 4

teaching (presenting ideas and information) and you are responsible for learning. It is expected that as an adult, you will take charge of your education. In college you need to evaluate your responsibilities and manage your time wisely. College instructors will not check to see if you have done your homework or if it is turned in on time. That is all up to you. You have to read all assigned work whether it is covered in class or not. Your may find that instructors may test you on information never covered in class.

Exercise 1.2

Will you be able to handle the differences between high school and college? In the space below, explore with your instructor and classmates the differences you may encounter and how to deal with them.

Recommended website on high school & college:
`http://www,about`
`college.com`

_____

_____

_____

_____

_____

_____

_____

Learning Styles and Study Strategies

Now let's turn to learning styles and reading. Since you will be required to process an enormous amount of information, it would be helpful to look at the way people tend to process information. The way in which you take in the information and how you make sense of the information is referred to as your learning style. If you can identify your preferred method of processing information, this will help you to capitalize on your strengths. In addition, this will give you the advantage of looking for instructors with the same teaching style. Take the LSI now before preceding any further.

Learning Style Inventory (LSI)

Directions: The purpose of this inventory is to help you assess how you prefer to learn. Read each item carefully and then circle the number of the response which best describes you: (1) Almost Never; (2) Rarely; (3) Sometimes; (4) Frequently; and (5) Almost Always. This is not a test. There are no right or wrong answers. Be honest with yourself.

Responses   **Items**

1 2 3 4 5   1.   I like using my hands when learning about something.
1 2 3 4 5   2.   I like seeing how a task is done before I try it.
1 2 3 4 5   3.   I would rather learn about the news listening to the radio than reading about it.
1 2 3 4 5   4.   I listen to the tone of the speaker's voice for the meaning.
1 2 3 4 5   5.   When someone is talking, I get a lot out of how that person uses gestures.
1 2 3 4 5   6.   When people introduce themselves, I try to visualize their names.
1 2 3 4 5   7.   I would rather participate in an activity than watch others do it.
1 2 3 4 5   8.   I tune a radio more by sound than by the numbers on the dial.
1 2 3 4 5   9.   When someone gives me a complicated problem, I prefer to see it on paper rather than hear about it.
1 2 3 4 5   10. I prefer classes in which I am actively doing something.
1 2 3 4 5   11. For me doing is learning.
1 2 3 4 5   12. For me seeing is believing.
1 2 3 4 5   13. I choose my clothes by the way they feel on me.
1 2 3 4 5   14. I visualize events, places, and people.
1 2 3 4 5   15. I depend upon the radio for keeping up with what's happening in the world.
1 2 3 4 5   16. I like to explore objects by feeling their texture, shape, and so forth.
1 2 3 4 5   17. I would rather watch an athletic event than participate in it.
1 2 3 4 5   18. I learn best when I can discuss my ideas with others.
1 2 3 4 5   19. I learn best from teachers who have distinctive voices, speaking, and lecturing styles.
1 2 3 4 5   20. I can tell more about persons from hearing their voices than from seeing them.
1 2 3 4 5   21. I like reading books which are illustrated better than books with no pictures or graphs.
1 2 3 4 5   22. I enjoy classes in which the teachers use many visual aids.
1 2 3 4 5   23. I would rather listen to records than read.
1 2 3 4 5   24. I like classes that have planned activities and experiments.
1 2 3 4 5   25. I am fascinated by sounds.
1 2 3 4 5   26. Listening to music is one of my favorite pastimes.
1 2 3 4 5   27. The things I remember best are the things in which I have participated.
1 2 3 4 5   28. In trying to remember where I left something, I visualize where I placed it.
1 2 3 4 5   29. Learning only has meaning for me if I get a chance to try it.
1 2 3 4 5   30. When learning, I like to sit back, listen, and absorb what is being said.

SCORING

To derive your scores, simply add the values for the items on each scale.

Visual Learning		Kinesthetic Learning	
2	_____	1	_____
5	_____	7	_____
6	_____	10	_____
9	_____	11	_____
12	_____	13	_____
14	_____	16	_____
17	_____	18	_____
21	_____	24	_____
22	_____	27	_____
28	_____	29	_____
=	_____	=	_____

Auditory Learning	
3	_____
4	_____
8	_____
15	_____
19	_____
20	_____
23	_____
25	_____
26	_____
30	_____
=	_____

A person's learning style preferences can be graphed in the following manner. For each learning mode, draw a line at the appropriate level and then complete the bar with a pencil.

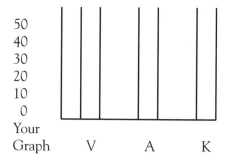

50
40
30
20
10
0
Your
Graph   V   A   K

DISCUSSION

Each of us develops preferred ways of learning. Some like to learn by watching others and observing how a task is done. Some prefer to learn by listening and discussing a topic. Others prefer to learn by doing or by trying things out. We call these three preferences—**visual**, **auditory**, and **kinesthetic**. You can discover your relative preferences by adding the values you circled according to the scoring keys.

Visual Learners: Those who have a preference for learning through the visual mode like to learn by reading, watching, and observing. They like to visualize when they learn. They learn best when they can see how things are done or how things are related. Visual aids—movies, film strips, pictures, graphs, and so forth—help visual learners. Visual learners are usually good spellers because they visualize words and can tell from the way a word looks if it is spelled correctly.

Auditor Learners: Those who have a preference for learning through the auditory mode like to learn by listening to lectures and discussions. They learn best when they participate in discussions or respond to questions. Auditory learners like to listen to tape recordings of material and have a chance to ask questions about what they have learned or do not understand. Seminars and discussion classes fit auditory learners best. Auditory learners usually have some difficulty in spelling because they try to spell by sounding words out; but, many of the same phonetic sounds in English can be spelled in a variety of ways.

Kinesthetic Learners: Those who have a preference for learning through the kinesthetic mode like to learn by doing. They often find all the explanations, visual presentations, and discussions confusing. Kinesthetic learners prefer to use a trial and error method of learning. They have strong feelings as to whether something is right or wrong, good or bad, but often have difficulty explaining why they feel that way. Kinesthetic learners do well in classes where they can use their hands to create and develop what they learn (classes like woodworking, cooking, typing, experiments, etc.). Kinesthetic learners usually have some difficulty in spelling.

SUMMARY

Knowledge of your learning preferences can be used in structuring your learning experiences and studying. Selection of classes, choice of teachers, and selection of subject matter can be enhanced with this knowledge. Learn to build on your preferences to increase your learning effectiveness. You can use this information to develop and strengthen other learning modes.

Reading is very important in college. You will quickly find that it is not just recognizing words but using critical thinking and getting the intended message of the writer. Reading for a lot of freshmen students is a difficult task simply because you don't like reading text books and have never had to read so much in such a short amount of time. College texts have more concepts and ideas on one page than most high school books. Therefore you need strategies that are going to keep you actively involved and focused. One of the reading strategies that's going to keep you active is the **SQ3R** five step method developed by Francis P. Robinson in 1941. It is the most widely used study strategy that has been proven to work. After working with college students with families and/or who were working full time, some of them still needed more help. The solution was the **PSQ3R**. Actually, its an addition to Robinson's strategy. Know the Purpose (P) for which you are reading before you start. If you have questions to answer, read them first. If you have objectives to learn, read them first.

P   Know the purpose for which you are reading.

S   Survey or skim the chapter before you read it. Also read the introduction and summary.

Q   As you move through the chapter, turn headings into questions. Now read to answer the questions.

R   Read with the purpose of answering questions you have generated from the questions. Read to answer questions the instructor gave you. Read to understand objectives. Mark up your text and write notes.

R   Once you have answered the questions and have completed each section of reading, recite aloud in your own words the answers.

R   After reading the entire chapter, review all information. Close your book and see what you can remember by reciting.

If you are taking a course that requires you to learn a skill, you need to add another "**P**" at the end which means practice. A skill cannot be learned unless it is practiced over and over. Mathematics is an example of one of the courses that requires you to learn skills through practicing it over and over.

Recommended Math Study Skills websites:
http://www.mathpower.com
http://www.iss.stthomas.edu/studygs.net/attmot4.htm

Tips on Keys to Excellence

Students learn survival skills

Tip 1:   Get to know your instructors.

Get to know them firsthand and form a positive relationship. Form your own opinions about them.

Tip 2:   Get to know your instructors by name.

Make sure they know your name by introducing yourself not only in the classroom but also on breaks and during their office hours. Make yourself an assignment to meet your instructors by the second week of class at the latest. Find out how the instructors wish to be addressed.

Tip 3:   Always sit in the front of the classroom.

In a traditional classroom, always sit in the front. This helps you focus on what is being presented without distractions. Besides, the front half of the class tends to make higher grades.

Tip 4:   Arrive early for class.

You can visit with your instructors or classmates, review notes, and/or spend a few minutes relaxing. Being on time demonstrates your commitment and interest. Do not leave early unless you have an emergency. In on-line courses, set up a weekly schedule and stay with it.

Tip 5:   Be prepared to take notes.

Buy and use a **different 8½″ x11″ notebook** (tablet) for note-taking in each class. Take notes.

CALVIN & HOBBES © 1992 Watterson. Dist. by UNIVERSAL PRESS SYNDICATE. Reprinted with permission. All right reserved.

Tip 6:   Participate in class discussions.

Ask questions. Provide answers. Be ready to debate and discuss ideas in class. Pay attention. This demonstrates to your instructors that you are interested and prepared. However, do not ask questions just to sidetrack your teachers or to get noticed. This only wastes everyone's time.

Tip 7:   Use the rules for communicating in the traditional classroom.

Raise your hand if you have a question or a comment unless you have the instructor's attention already. Converse only with the instructor when lectures are being presented.

Daydreaming, sleeping or having side conversations will insult your instructor. Besides, you miss what is happening and disturb other students. Avoid swearing in the classroom. It is considered unacceptable behavior.

Tip 8:   Communicating in on-line courses

If you are taking on-line courses, find out how the professor wants to be addressed and use proper etiquette for sending e-mails, chatting, and using the discussion board. Never type in all capital letters when sending e-mails, discussing or when using the chat room.

Tip 9:   Know your instructors' expectations.

During the first class, find out what is expected of you. Ask questions such as

- What are the objectives of the course?

- What assignments do I have to complete?

- What is the grading policy?

- When will examinations be given?

- What materials/chapters will examinations cover?

- What are faculty office hours?

Tip 10: Use an academic calendar for recording class times and assignment due-dates.

Also, note time for studying and test dates. The calendar can usually be bought in the campus bookstore.

Tip 11: Cheating and plagiarism are not accepted in college

Beginning students frequently use someone else's material without giving credit. This is unacceptable behavior in college. If you use an author's work, then give credit by citing the material. See chapter three on "Searching for Information."

Tip 12: Plan to complete all assignments on time.

Avoid excuses and/or procrastinating. Instructors know most of them. Most teachers can see a snow-job coming before you can finish thinking it up. Don't forget that your teachers were once students, too.

Tip 13: Always turn in professional work of high quality in both content and form.

Better-looking papers tend to get higher grades. In preparing your paper, work on it as though it would determine whether you will get a raise or a promotion. Neat papers of high quality demonstrate your interest in the course. The use of a word processor to complete assignments is highly recommended. Learn the rules for formatting information.

Tip 14: Begin preparing for exams from the first day of class.

Review notes as soon as possible after every class.

Tip 15: Know what's going on in class.

If you don't, make an appointment to see your teachers during office hours. Most instructors are delighted to see you and will go out of their way to assist you. If you are taking on-line courses, try to set up a time when you and the instructor can meet in the chat room so you can discuss areas of study in which you may be having problems.

Tip 16: Use positive conflict management.

Resolve differences of opinion by calmly sitting down and discussing the issue with your teachers. Remember, everyone has a right to his/her opinion, but in the classroom, the last word belongs to your teachers.

Sometimes you may not be able to resolve your differences on the spot. When this is the case, make an appointment to see your instructor during his/her office hours so you do not hold up or disrupt that day's lesson. Follow these steps for solving differences with your instructors.

- Level 1   Go to your *instructor* and explain your problem. If you have not been able to resolve it, move to the next level.

- Level 2   Go to the *department chairperson* and explain your problem. Include how you tried to resolve it with your teacher. Finally, if the problem is still unresolved, move on to the last level.

- Level 3   Go to the *Associate Dean* or the *Director of the division*. Explain your problem and what has transpired with your teacher and the chairperson.

Tip 17: Attend class.

Always come to class. If you are absent as a result of an emergency, be sure to check with your instructors about your assignments (or what you missed in class) before returning to class.

Coming to class is extremely important in courses such as mathematics and English where you must learn a skill and information builds on previous information.

Most on-line courses have assignments that require discussions in a timely manner so that if you are not on-line weekly, you will be penalized for not participating.

Tip 18: Buy your books early.

Skim through each book and get the general impression of the main ideas and how they are presented. Don't wait for an assignment before opening your text.

Tip 19: Begin observing the instructor's style of teaching during the first class.

This will help you figure out the importance of information being presented and what will probably be on your tests.

Tip 20: Self-discipline is your best friend.

This bit of information will make all the other tips work for you. It requires that you recognize the role of sacrificing in order to achieve. You do have some say in what happens to you. You must want to excel if you want to achieve.

Tip 21: Keep an open mind and a positive attitude.

You will encounter some obstacles along the way to graduation, but you must simply take them as challenges on the way to starting your career.

Tip 22: Keep an open mind to cultural differences.

You will encounter teachers and peers who look different, act differently, sound different and have differing points of view. Consider these differences as a new learning experience and a chance to learn more about other cultures.

Tip 23: Focus early on a career choice.

You need to decide on a career now. All academic planning is based on your career goal. An advisor is not able to effectively help you select courses if you don't know what you are planning to do.

Tip 24: Become involved; join a club.

Form a study group. Exchange phone numbers with other students in your classes with whom you think you can work. Students who are involved and actively participate are less likely to drop out of school. Make the move and better your chances for succeeding.

On-line students need to exchange e-mail addresses and form chat and/or discussion groups.

Tip 25: Be considerate of other students.

Avoid impolitely groaning and/or yawning aloud. Do not wear hats and other headgear that may obstruct the view of other students sitting behind you. Avoid putting feet on top of desks. Avoid blocking the view of students watching visual-aids in the classroom.

If you happen to be an on-line student, be careful how you say what you say. When you put something in writing, try to view if from the other person's eyes before sending it.

When discussing, remember to comment only on the content of what is being discussed.

Tip 26: Don't use personal electronic equipment in the classroom.

Avoid using beepers, alarms, and/or telephones in the classroom. Do not allow them to go off and disturb the learning process. You may want to turn them off until your class is over.

Tip 27: Learn to use computers for academic success.

Today, as in the future, you will be required to use the computer. If you don't have a personal computer at home, use ones provided for you at the college to complete homework and do research. Your instructor will give you many options for communicating with them. Some of them may include using the Internet, e-mail and the FAX. Learn to use them now.

Making the Grade

Making the grade in college is important because setting your sights on academic excellence is one of the smartest things you'll ever do.

The better your academic record, the better your chance to

compete successfully;

graduate on time;

get a scholarship;

transfer to a good school;

get a job.

Working hard to achieve academic excellence will help *you develop skills* that employers are looking for and *build confidence in yourself*.

Exercise 1.3

1. In a small group, discuss five tips you already know work for you. Now discuss another five you had not given much thought to and how they might work for you. Get some help from others in the group to complete this exercise.

2. Visit with one of your instructors, introduce yourself and ask him/her for tips-on passing his/her class and report back in the next session. Try to get to know your instructor.

3. Class participation is extremely important in a number of classes and actually counts as part of your grade. Devise a strategy for yourself and others who may not be used to sharing and/or asking questions in class as to how it can be done.

4. Introduce yourself and share with your classmates why you believe you are going to be successful in college.

? **Journal Questions/Activities**

1.   Identify your learning style.

2.   Now that you have had some time to think about your learning style, do you believe it is accurate? Explain why or why not.

3.   What kind of changes do you believe you will need to make given all the information provided in this chapter if you are going to be a successful student. Be specific.

4.   Are you comfortable with where you are and who you are?

Summary

The first part of the chapter explained the role of colleges, the many tasks of faculty, listed some of the characteristics of educated people and identified some of the skills you will need as students to survive in college. These skills include:

Organization; time management, goal setting, note-taking, taking tests (effective use of study skills); being motivated; and the sensitivity to other cultures in the environment.

The chapter also provided you with information on your learning style, discussed the transition from high school to college and provided a myriad of tips for getting started on the right track in college. If you overlooked the twenty-five tips, go back now and read them. They include information on how to communicate with your instructor and how to deal with situations in the academic environment.

Summary Exercise 1.4

Chapter Test. As a final test of your understanding of this chapter, answer true or false to the following items:

    T   F

1. ☐ ☐   College is a teaching environment.

2. ☐ ☐   Educated people are said to be non-reliant.

3. ☐ ☐   Educated people are considered to be responsible for their actions.

4. ☐ ☐   The faculty's major task is to assist you in learning the ABCs of life.

5. ☐ ☐   A syllabus outlines your rights as a student.

6. ☐ ☐   If you are having a problem with an instructor, your first step is to go to the chairperson of the department before the situation gets out-of-hand.

7. ☐ ☐   Successful students rarely, if ever, miss class.

8. ☐ ☐   Unsuccessful students tend to miss many sessions and never try to make up the time or discover what work they missed.

9. ☐ ☐   Self-discipline simply means coming to class on-time.

10. ☐ ☐   Good grades and club involvement tend not to have any effect on one's career.

References

American Association of Community Colleges. "The Commmunity College Impact." Trends and Statistics, Phillippe & Patton, 2000. `http://www.aacc.nche.edu`. 2004.

Feldman, Robert S. POWER Learning: Strategies for Success in College and Life, 2nd edition, New York: McGraw-Hill, 2003.

Freeman, Ellen. Professor Freeman's Math Help. 2003 `http://www.mathpower.com`.

Herlin, Wayne and Mayfield, Craig. *Successful Study Skills*, Dubuque, Iowa: Kendall/Hunt Publishing Co., 1981.

`http://factfinder.census.gov`. 2004.

Landsberger, Joe. The Study Guides and Strategies websites, University of St. Thomas, St. Paul Minnesota. 2003. `http://www.iss.stthomas.edu/studygs.net/attmot4.htm`.

Miami Dade College. `http://www.mdc.edu`.

Scriptographic Booklet—*About Making the Grade*, South Deerfield, MA: Channing L. Bete Co., Inc., 1984.

Walter, Tim and Siebert, Al. *Student Success: How to Succeed in College and Still Have Time for Your Friends*, 5th ed., Harcourt Brace Jovanovich, Inc., 1990.

Wayne State University. UGE 100: *The University and Its Libraries—Readings*. Detroit, Michigan: Wayne State.

Chapter Two

Campus Resources

Sheryl M. Hartman, Ph.D.

Exercise 2.1

Campus Resources Awareness Check

DIRECTIONS: Please place an "X" in the appropriate box.

	Yes	No	
1.	☐	☐	You can obtain a transcript in the registrar's office.
2.	☐	☐	You can copy a tape and view a film in the Student Activities office.
3.	☐	☐	You have to pay to use the athletic facilities on campus.
4.	☐	☐	You must call the campus operator to find your instructor's office.
5.	☐	☐	The only performing group on campus is the marching band.
6.	☐	☐	The students' rights are listed in the catalog.
7.	☐	☐	No tutors are available on campus.
8.	☐	☐	Student organizations are limited in number and membership.
9.	☐	☐	You can obtain help for personal problems in the counseling office.
10.	☐	☐	The Student Employment, Career and Transfer Center has information on jobs for students online.

Introduction

Academic environments provide many resources and services. Professors, lab personnel, administrators and other key personnel are academic resources who teach, challenge and mentor students while encouraging academic excellence and the development of critical thinking and problem-solving skills. High quality academic training is delivered via the use of state-of-the-art technology, accessibility to academic materials and training expereiences, and participation in the college environment.

Services for Students

ADMISSIONS/REGISTRATION

The College Admissions Office processes applications for admission. Students apply here for college credit programs and continuing education programs.

The Office of the Registrar is the custodian of all official student academic records and maintains student information such as current address and chosen program of study. It provides both official and unofficial copies of student academic records to students or to other individuals, institutions or agencies upon request from students.

Class schedules are produced and transcripts are issued from this office. All changes to the students' academic schedule, such as adding and dropping courses and withdrawing from the college are processed here. Students should print or pick up a new copy of their schedules following these changes. Students are encouraged to notify the Registrar of any corrections to their personal information.

Registration for courses is held each term according to scheduled dates. Students may register in person, on the Internet, or via the telephone.

FINANCIAL AID

Financial aid packages consisting of grants, scholarships, loans and college work study are available to support a college education. Students may now apply for financial aid through the Internet. It takes approximately ten days for these applications to be processed.

BURSAR'S OFFICE

This is where students pay their fees or process their final financial aid papers after scheduling their classes. Fees may be paid in cash, by check or via web site.

TESTING

College credit-seeking students receive assessment testing to ensure enrollment in the proper classes. The Testing Office administers the College Placement Test (CPT) or other comparable assessment tools. Additional tests administered include the:

Scholastic Aptitude Test (SAT)
`http://www.collegeboard.com/student/testing/sat/about/aboutFAQ.html`

American College Test (ACT) `http://www.act.org/aap/`

Test of English as. a Second Language (TOEFL)
`http://www.toefl.org/`

College Level Entrance Program Exams (CLEP)
`http://www.collegeboard.com/student/testing/clep/exams.html`

College Level Academic Skills Test (CLAST)
http://www.firn.edu/doe/sas/clsthome.htm

ACADEMIC ADVISEMENT

All new students are guided in their initial college enrollment by an academic advisor. Following initial admission to College and assessment testing, students begin to develop an educational plan. The advisor assists them in this task, advising students on the courses they will register for during their first academic term. Advisors are available on a walk-in basis, by appointment, via telephone, and over the Internet. Students are encouraged to understand their educational degree requirements and to project their academic course responsibilities.

General advisors analyze the student's completion of common education requirements, recommend elective courses based on career choices, and check graduation eligibility. When a student becomes established in the academic environment and has declared a major, they should seek academic advisement from faculty in their major department.

Many colleges now provide a computerized degree audit system, which will track a student's progress toward meeting graduation requirements. These online tools may also help a student to compare educational institutions, and will document educational benchmarks the student completes.

Advisor explains how to select courses for major

STUDENT CAREER CENTER

State-of-the-art technology is available to assist students in their career search, helping them to make decisions about interim and lifetime career planning. Accessible career planning software includes some of the following tools:

SIGI PLUS Educational and Career Planning Software

- Assesses your career interests, abilities and values

- Lists schools and colleges that offer training
  http://www.ets.org/sigi/

Choices

- This is a tool that helps you to expand the occupational fields you explore. The computerized inventory identifies occupations based on your preferences for:
  Physical demands
  Education level
  Temperament
  Salary requirements
  Aptitudes
  Career fields
  Work environment

Occupational Outlook Handbook

- Provides basic information about occupation, nature of work, salary, job market, training requirement
  `http://www.bls.gov/oco/`

The Myers Briggs Type Indicator (MBTI)

- A well-used indicator of adult personality patterns

- The results of the test can be used to help guide decisions regarding majors or careers, as well as to help understand and appreciate individual differences in interpersonal relationships.

- Many educational institutions offer this inventory with professional interpretation for students.

STUDENT EMPLOYMENT

Most student employment centers now offer career seminars and workshops focusing on resume writing and interviewing skills. These centers also serve as a referral site for student employment, including work study positions, part-time employment, full-time employment, and internships. Many career centers maintain a job bank with daily updating of listings. They also sponsor on site interviews with national and area recruiters and other activities which promote employment activities, such as Job Fairs.

DISABLED STUDENT SERVICES

Students with disabilities have specific rights that are guaranteed to them by law.

The Americans with Disabilities Home Page can be reviewed at the following location, `http://www.usdoj.gov/crt/ada/adahom1.htm`. This Act provides you with the following rights:

- You have the right to ask every instructor to provided either modifications or accommodations for you that your disability requires. This may include extra time in testing, an alternate testing environment, and academic material provided in various formats.

- You may tape record your classes, so long as you guarantee that the tapes you make will only be used by you and not shared with anyone else.

Access services supporting disabled services may include student readers and note takers.

BOOKSTORE

The best way to purchase the correct books for your classes is to visit your college bookstore. Most purchases can now be completed via the Internet. College bookstores provide new and used textbooks, course

related study guides, educationally priced software, college gear and personal items.

LIBRARY

College libraries make available extensive electronic information from web-based databases These include general access to ERIC (Educational Resources Information Center); First Search—subject access to citations of journal articles, government publications, and conference proceedings on education, engineering, medicine, law, and other subjects; WorldCat— database of bibliographic information including over 30 million books and non-print materials; and the Grolier Academic American Encyclopedia. Also available are collections of microfilm, periodicals, reference books, newspapers, magazines, and audio-visuals.

AUDIOVISUAL SERVICES

Audiovisual services provide access to tape, film, and video equipment. Adhering to copyright legislation, students may copy cassettes and tapes for classes such as court reporting, view films and audiovisual material, and record sound.

INTERNATIONAL STUDENT ADVISOR

International students may obtain information on housing, visas, and registration for classes.

Student Life

Student Life offers students a chance to enhance their educational experience through cultural, social, and educational events.

STUDENT ORGANIZATIONS

The following list includes a sampling of the many student organizations catering to a variety of student interests.

African American Student Union
American Institute of Architects
American Sign Language
Association of Student Engineers
Black Student Union
Caribbean Student Association
Catholic Campus Ministry
Circle K International
Criminal Justice Service Club
Ecology Club
Falcon Cheerleaders
Falcon Times
Future Teachers of America
GEPAF Group Etydian Progresis Aysien Nan Florid
Hillel Jewish Student Union
International Language Club

LASO—Latin American Student Association
Mass Media Association
Out on North
Phi Beta Lambda
Phi Lambda
Phi Lambda Phi
Phi Theta Kappa
Physical Therapy Majors
Programming Advisory Board
Science Club
STARS—Students that are Ready to Serve
Student Government
Travel Club
WAVES

Students who do not find a club to match their interests may often start their own organization by contacting the Student Life Department.

Performing Arts

Performing arts organizations include music, drama, and dance groups and are open to all students who have an interest or talent in playing an instrument, singing, acting, dancing, or providing technical assistance. The directors of the groups are particularly interested in attracting students who participated in high school bands, orchestras, choirs, dance groups, or drama clubs. Students do not have to be music, drama, or dance majors to participate.

Athletics

INTERCOLLEGIATE ATHLETICS

The purpose of intercollegiate athletics is to promote excellence in athletic participation for men and women athletes. Students interested in playing for a team should contact the appropriate coach.

PHYSICAL EDUCATION FACILITIES

Many recreational facilities are available to students any time classes are not using those facilities. These facilities may include a jogging track, fitness court, tennis courts and swimming pool.

COLLEGE CATALOG

The College Catalog contains historical information about the college, academic and admissions regulations, descriptions of programs and courses, and listings/educational backgrounds of all administrators and faculty.

STUDENTS' RIGHTS AND RESPONSIBILITIES

This handbook describes topics pertinent to student success such as evaluation criteria, student records, student affairs, and discipline. Policies on student behavior as well as teaching/learning values are detailed.

CONCLUSION

The goal of the higher education environment is to provide the learner with strategies to enjoy a lifetime of learning. There are many resources to assist the serious student in accomplishing their educational goals.

Exercise 2.2

1. Review the Miami Dade College wide website. `http://www.mdc.edu`.

2. Look at the academic calendar. Identify when the academic term ends, and when registration for the next term begins.

3. Review the course catalogue, and read the descriptions of two elective courses you will need to take.

4. Read the items of interest that have been sent to local media in the last month. (Hint—look at news stories).

5. Visit Florida's Official Online Student Advising System `http://www.facts.org` and conduct some research on your college choices.

❓ Journal Questions/Activities

1. What resources on campus are most important to you?

2. What resources do you need that you do not find on campus?

3. Identify two campus resources you will be investigating further.

Summary Exercise 2.3

True/False Questions

1. Financial aid packages may consist of grants, scholarships, loans and college work study assignments.

2. The Office of the Registrar is the custodian of all official student academic records.

3. Student employment centers offer career seminars and workshops focusing on resume writing and interviewing skills.

4. Students with disabilities have specific rights guaranteed to them by law.

5. Choices is a career assessment tool that helps you to expand the career options that you explore.

Multiple-Choice Questions

1. You may tape record your class:
   a. as long as you don't charge more than $10 per tape when you sell the tape.
   b. turn on soft music to help student's study when they listen to the tape.
   c. guarantee that the tapes you make will only be used by you and not shared with anyone else.
   d. make the taped information accessible to all of your classmates.

2. The Myers-Briggs Type Indicator
   a. demonstrates your interest in what the instructor is saying.
   b. is a well-used indicator of adult personality patterns.
   c. Both a and b.
   d. Neither a nor b.

3. The SIGI PLUS Educational and Career Planning Software:
   a. Assesses your career interests, abilities and values.
   b. Lists schools and colleges that offer training.
   c. Both a and b.
   d. Neither a nor b.

4. Academic advisors:
   a. recommend elective courses based on career choices.
   b. analyze the student's completion of common education requirements.
   c. Both a and b.
   d. Neither a nor b.

5.   Students with disabilities have the right to ask every instructor to provide the following accommodations or modifications:
     a.   An alternate testing environment.
     b.   Extra time in testing.
     c.   Both a and b.
     d.   Neither a nor b.

Fill in the blank

1.   Tests such as the College Level Entrance Program exams (CLEP) and the Scholastic Aptitude Test are administered in the _____ Office.

2.   The Office of the _____ maintains student information such as current address and completion of degree.

3.   It takes approximately 10 days for financial aid applications to be processed when they are submitted via the _____ .

4.   An academic _____ will guide students in their initial selection of college classes.

5.   Sources of electronic information such as ERIC (Educational Resources Information Center) and First Search (subject access to citations of journal articles, government publications, and conference proceedings on education, engineering, medicine, law and other subjects) are known as _____ .

Chapter Three

Searching for Information

Georgina Fernández, M.L.S.

Name _____   Date _____

Exercise 3.1

Searching for Information Awareness Check

DIRECTIONS: Please place an "X" in the appropriate box.

	Yes	No	
1.	☐	☐	The literature of a field is the body of written work produced by researchers or scholars.
2.	☐	☐	There is only one method of finding publications.
3.	☐	☐	A bibliography is a list of books and articles on a topic.
4.	☐	☐	A citation is a short way of identifying a publication.
5.	☐	☐	It is legal to omit the references for the books and articles you used for your paper.
6.	☐	☐	Searching the literature of a field (books and articles) is part of scholarly study.
7.	☐	☐	Critical thinking skills will help you in searching for information.
8.	☐	☐	The first step in searching the literature of a field is to prepare a list of materials you are going to use.
9.	☐	☐	Reference librarians can assist you in finding the library materials you need.
10.	☐	☐	You can find important information through the library's online systems.

Introduction

In the face of technological advances and the increasing flood of information, we are constantly challenged with one basic question. What is the best choice? This chapter focuses on the need for students to make appropriate choices when presented with the numerous available sources of information. You will learn principles that are essential in the process of searching for information. You will be able to produce a list of materials on a topic of interest by using a proper search strategy. Lastly, you will become familiar with the organization, capabilities and services of the Miami-Dade College libraries.

Ways to Information

The literature of any field of study contains publications such as books, articles and reports from various sources which form a highly organized system for communicating information. Reviewing the literature that deals with the focus of your interest is an excellent way of learning about a subject. Of course, it is crucial that you locate publications that specifically address your topic. For example, to find pertinent publications on the impact of illiteracy in the United States in the 2000s, you will need to search through the publications in the field of education.

Methods of Locating Publications

Students and researchers employ three methods of finding publications in their subject of interest.

The first method consists of looking up the references cited in books and articles. For example, you can select cited publications from a textbook or an encyclopedia article. This method has three disadvantages. First, it may provide books and articles with only a narrow focus on the subject under consideration; second, it may fail to introduce the broader viewpoints of other writers; and third, it does not contain references that have been published after the book or article you are reading.

Cited references

The second method of finding publications relies on personal familiarity with what has been written on the subject. It is easy for scholars to use this method because they are immersed in the study of their subjects and can draw from their extensive knowledge and experience. On the other hand, this method is difficult for students because they have a limited familiarity with the publications in a field. Even scholars may find situations in which a new question forces them to become acquainted with an unfamiliar area of the publications on their subject.

Personal familiarity

A third method of finding information in the literature is through the use of "research tools." These are print publications or electronic resources that tell you where to go for information and serve as tools or **indexes** to

Research tools

locate sources relevant to the subject under study. With a basic understanding of the different types of tools you will be able to locate the information you need. For example, PsycFIRST is one of the several online databases or research tools available to you as a Miami-Dade Community College registered student. This online database will help you identify full-text articles in psychology, social work, sociology and other related fields. This method of locating publications will be discussed further in "Your Library" on page 47.

What Is a Bibliography?

A bibliography represents the available publications on a specific subject, a more or less comprehensive list of books and articles on a topic. A bibliography can be compared to a map in that it can lead you to places where you can find the information. Bibliographies may represent various aspects of information about a subject. For example, a bibliography on Jews as a subject may comprise a list of books on Jews in the American Civil War, Jews on civil rights organizations, Jews as slave owners, Jews as merchants, etc.

The following bibliography represents various books and periodical articles on academic success of college students.

COLLEGE STUDENT ACADEMIC SUCCESS: A SELECTED BIBLIOGRAPHY

Articles for Academic Success. Florida Institute of Technology. www.fit.edu/caps/articles/academic_success.html. Dec 2003/

Bishop, Joyce, Mary Jane Bradbury, and Julie Wheeler. *Keys to Success Reader.* Upper Saddle River, N.J.: Prentice Hall, 1999.

Combs, Patrick. *Major in Success: Make College Easier, Fire Up Your Dreams, and Get a Very Cool Job.* Berkeley, CA: Ten Speed Press, 2000.

Feldman, Robert. *P.O.W.E.R. Learning: Strategies for Success in College and Life.* Boston: McGraw-Hill, 2000.

Gardner, John N., A. Jerome Jewler. *Your College Experience: Strategies for Success.* Belmont, CA: Wadsworth, 2000.

Montgomery, Rhonda J., Patricia G. Moody, and Robert M. Sherfield. *Cornerstone: Building on Your Best.* Upper Saddle River, N.J.: Prentice Hall, 2000.

Osborne, Jason W. "Identification with Academics and Academic Success among Community College Students." *Community College Review.* 25 (1997) 59-67.

Tips for Academic Success. http://www.stthomas.edu/academiccounseling/Tips.htm.

Works Cited List/Plagiarism

You will always need to present a list of the works consulted, at the end of any paper which has made use of outside sources. In other words, it will be a list of references to the publications you read. If you fail to acknowledge the use of a work, you are committing **plagiarism**, the stealing of another author's thoughts or intellectual property. And if you get caught plagiarizing you could be dismissed from the college and could even face legal charges.

You must present each publication in a short format, or citation, which is a way of identifying the publication used. For books, this includes author, title, place of publication, publisher, and copyright date.

Figure 1 illustrates the manner in which a book citation is presented.

Sternberg, Robert J. *Successful Intelligence: How Practical and Creative Intelligence Determine Success in Life.* New York, N.Y.: Simon and Schuster, **Fig. 1.** 1999.

Table 1.0. Examples of Citations for Different Types of Sources

Book with one author	Dawson, Lallie P. *Bigotry and Violence on American College Campuses.* Washington, D.C.: United States Commission on Civil Rights, 1990.
Book with two or more authors	Nist, Sherrie L. and Jodi Patrick Holschuh. *College Success Strategies.* Lebanon, IN: Longman Publishing Group, 2002.
Article in a scholarly journal	Donlevy, James G. "Reaching Higher Standards: Special Education, Real-World Certification in Technology and the Community College Connection." *International Journal of Instructional Media.* 26 (1999) 241-8.
Article in a magazine or newspaper	Mariano, Carmen. "Smart Enough to Excel." *Vital Speeches of the Day.* 1 Nov. 1999: 62.
Article with no author names	"Miami-Dade Women Accepted into Delta Psi." *Miami Herald.* 28 Dec. 2000: E13.
Article in a database	"Need for Pay Packet High Achievement." *Times Higher Education Supplement* 14 May 1999: *2 EBSCOhost Academic Search Premier.* `http://www.ehostvgw8.epnet.com.` 17 Mar 2001.
Internet site with author	Ivey, Keith C. "Untangling the Web: Citing Internet Sources." `www.eeicommunications.com/eye/utw/96aug.html.` Aug 2003. Accessed 05 January 2004.
Video recording	*Building on Your Best.* Videocassette. Boston, Mass.: Allyn and Bacon, 1997.
Interview	Suárez, Celia, Ed.D. Personal interview. 28 March 2001.

Focusing Your Research

The process of searching material for publications on your topic is an essential part of any scholarly study and should be approached systematically. Students often assume that the best way to start a research paper is to gather the entire bibliography at the beginning, thus easing the task of reading and writing. This approach is incorrect because the materials

Literature searching is part of scholarly study

chosen at this point are done without your full **understanding of the subject**. When you start an investigation, you are relatively unfamiliar with the area of study. Although you may have some knowledge of the subject, you are unlikely to have specific questions because your understanding of the subject is not well developed. In summary, you do not have the background to choose from the references available to you. Without a good basic knowledge of the subject, you might end up choosing any publication that seems to refer to a term or phrase linked to the subject. Choosing your sources is like putting together an intelligent menu. In the same way that you prepare a balanced meal, you must select books and articles that reflect your understanding of the subject matter. The same errors of selection occur when students use the familiar way of starting the search at the library catalog. The problem here is that the catalog, regardless of its format, is only an inventory of the total contents of the library (or libraries). When you look up a subject in a library catalog, you are most likely to find a range of books that you can reasonably read. How do you know which of these books provide the information you need? You can only rely on clues such as titles which may mean little to you because you are not familiar with the subject. Your best approach is to concentrate on defining the focus of your paper and begin your search for relevant books and articles with the first step below.

Library catalog

LITERATURE SEARCHING

The **first step** consists of finding and reading a background article on your subject. For example, to prepare for a term paper, a good start might be to begin reading an article from an encyclopedia. Encyclopedias are the best examples of reference books that present an overview of a subject. Encyclopedias usually provide the general background of a subject and an up-to-date summary of current knowledge. For example, a work such as the *International Encyclopedia of the Social Sciences* can show the different ways of treating a topic within several branches of knowledge called disciplines. These divisions will guide you to the books and articles of that field that relate closely to your topic. For example, take the subject "fascism," the philosophy or system of government that favors the ruling of the extreme right. Treating the subject from the discipline of psychology occurs when the material deals with the psychological analysis of personalities; from religion when the article discusses fascism in relation to the Christian democracy; from sociology when social status is the focus of the discussion, etc.

Starting the search

General treatments of a topic are useful when you have no firm ideas about a specific topic or problem for your paper, or you are unsure how to narrow a broad and complex topic. As you read the generalized account on your subject, you may develop ideas on narrow issues that you will be able to explore further for your paper and find references to more specific literature. You will move from general to specific publications—from encyclopedias to articles—until you reach the level of detail that will cover the question you have in mind. This is the basic process of searching for information in scholarly publications.

Understanding the concept of this system of communication will allow flexibility of use in similar situations. Also, mastery of this process of searching information will help you in any computer online database searching. You might have little knowledge of the library, but the current online systems are designed to provide linkages among your own vocabulary, the conceptual language of the subject, and the language of indexes. This first step in literature searching is discussed in more detail in Research Strategy.

RESEARCH STRATEGY

Research strategy is the organization of the various steps that you must follow in a logical and efficient order to solve a research problem. Based on what you have learned so far, you can clearly see two elements: a **system and a question.** The system contains the tools that organize information (indexes, collection of materials, Internet search engines, or network), the vocabulary and the limitations of the system. The question is the topic you want to develop. The search process is the link that joins the system and the question. You have also learned that understanding the process of searching the literature will help you solve your future information needs. You will become efficient in finding the right information on your topic if you understand and use this concept. Below are the steps needed in a thorough research strategy.

1. **Define the focus of your search and clarify unfamiliar terms.** You need to read a background article on your subject. For example, a good start for a paper on hurricanes is to read an encyclopedia article with a general view on the subject. You will find relevant information on your subject by accessing the Internet for works such as the online *Encyclopedia Britannica* (www.britannnica.com) and by visiting your library to   consult the *McGraw-Hill Encyclopedia of Science and Technology.*

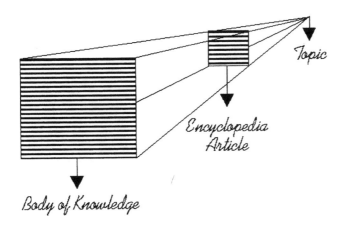

Fig. 2.

Steps for research strategy

2. **Break the topic into simple subtopics.** You might need to discuss your subject with your instructor or a librarian. It is necessary that you fully understand the special areas of a subject that you can consciously select. For example, since the subject "hurricane" is very broad, you can select the subtopic "motion of hurricanes."

3. **Estimate the quantity of material needed.** In most cases, your instructor will specify the number of items you need to consult for your finished paper. It never hurts, however, to jot down a tentative mixture and number of materials you might need. For example, you can anticipate if you should consult four to eight books, five to ten journal articles, and three to five newspaper accounts. The quantity issue will be determined by the type of research problem you have.

4. **Specify the quality of material.** The quality of material refers to the authority of its origin. Your instructors may or may not accept materials from the popular press such as *Time* magazine. They may require scholarly publications such as *Ecology* or they might want a combination of both.

5. **Budget your time.** You must determine at an early stage how much time is necessary for each part of the work. A good reason for this is that you might need to request materials from other libraries through the interlibrary loan department of your library which can delay your research up to several weeks.

6. **Identify types of research tools you will search. Research tools** do not always give you information directly, but they will lead you to the information contained in another book, periodical or Internet site. You will find several citations of articles on various aspects of your subject in several of the online databases available to students. You can access them from your personal computer in the comfort of your home by using your M-DCC borrower ID. Some of the databases give you just citations so you then need to locate the articles in your library or request them from a library outside of M-DCC through your interlibrary loan department. Other databases lead you to full-text articles that you can print at no cost. For example, the article "Sentinels in the Sky" by Jeff Rosenfeld published in *Weatherwise* of January-February 2000, is one of the relevant articles retrieved from one of the OCLCFirstSearch databases on hurricanes.

7. **Frame your question in the language of the system.** Your search words must match the subject headings of the system you use. For example, the search words used for the example presented above is "hurricanes" and "lightning."

Using your critical thinking skills to judge and select publications on your topic will help you develop answers to your questions. Surprisingly, with the exception of the first and sixth steps you can do all the others outside of the library. In many cases, you can adequately complete a thorough question analysis at home which in the long run saves you a great deal of time and frustration.

Your Library

All libraries contain collections of materials (books, magazines, journals, electronic book or e-books, records, microfilms, etc.) organized in a logical order and accessible through **indexes**, also called **research tools**. You can locate information sources through the use of those indexes available in your library and on the Internet. Generally speaking there are four major types of indexes or research tools.

Four major types of indexes

1. **The online library catalog** enables you to search for books and non-print materials owned by the college. You can also search the library catalogs of the other twenty-seven community colleges in Florida. In sum, the online catalog links the state's twenty-eight community colleges in one electronic system.

2. **Indexes to periodicals** list articles in magazines and journals but do not always provide the actual information. Indexes are available in print and non-print format. There are several online indexing services with or without full text in the M-DCC library resources. When you search a full-text online database you will first find a list of article citations relevant to your topic, then you can select the citations that will lead you to the full-text articles you want to retrieve. As you learned from previous pages, the *WilsonSelectPlus* is one of the several online databases with full-text articles available to you as a M-DCC registered student. You have access to all the online databases from any computer in the library or outside of the building. You can use any of the library's workstations located near the reference desk; they are dedicated to online search only. You can also use the computers housed in the Computer Courtyard on the first floor of the library building. To learn about the rules and regulations of the Courtyard you may visit its Web page at www.mdcc.edu/north/courtyard.

3. **Abstracting services** are identical to indexes to periodicals except that they include a brief summary of each article, so you can understand its contents rather than guess them from the title.

4. **Bibliographies** are lists of books and/or articles on specific subjects or subject areas. An example of a comprehensive bibliography in a book format is *Black-Jewish Relations in the United States 1752-1984: A Selected Bibliography*, compiled by Lenwood G. Davis, 1984. If you chose to write about the relations of black and Jewish people of this country, this book would be a marvelous tool for you. All you have to

do is identify the relevant citations to your topic, and then locate them in your library or through online services.

Example of a Bibliography in a book format:
A Research Tool

Fig 3.

The Internet

WHAT IS THE INTERNET?

Internet is a world-wide communications system that links computers together throughout the globe. In a sense, we could say that the Internet is the largest computer network, but in reality the Internet is formed by thousands of networks that exchange information freely.

Fig. 4.

The Internet serves as both research tool and comprehensive source of information; the appropriate use of the available Internet research tools or search engines (page 50) will lead you to the information you need. Using the Internet can be easy, moderately difficult or very difficult, depending on the scope of your research and the types of research tools you use. To use the Internet effectively you need to understand how it works.

THE WORLD WIDE WEB

Many people use the terms Internet and World Wide Web (WWW) interchangeably, but there is an important distinction between the two terms. The WWW is just a portion of the Internet, not even the largest portion. Basically, it is a system using a document formatting language that enables you to link documents in different computers that are connected to the Internet. The name of the language used is HTML (Hypertext Markup Language). This linking is a giant step toward improving the interaction among computers and people, and toward making the process of looking for information faster and easier.

Finding specific information in the Internet is like finding specific information in a library that lacks the research tools to locate sources of information. This is the reason the Web directories or Web sites are so useful; they are organized by topic to help narrow down your browsing, just like maps of cyberspace. The amazing fact about the WWW is that it creates connections between pieces of information from around the world that allow you to easily find related topics. Each Web site has an address, or Uniform Resource Locator (URL).

The URLs of the Web sites listed below are just examples of a wide variety of topics that you might want to check out or research. You can expand your list of useful Web sites by examining those listed on any of the M-DC Libraries Web Pages. The lists comprise URLs for Web sites that the M-DC librarians recommend as relevant and useful sources to support course assignments. Just click on the campus of your choice, and then on its Web page. For example, the URL for North Campus Library Web page is www.mdc.edu/north/library.

ASPCA
> www.aspca.com
>> American Society for the Prevention of Cruelty to Animals

Chateau of Versailles
> www.chateauversailles.fr
>> A most interesting art site from France

CNN Health
> www.cnn.com/health/index.html
>> General medical and health news

Fodor's Travel Online
> www.fodors.com
>> Trip planning and destination reviews

Freeman Institute
> www.freemaninstitute.com/diversity/htm
>> Diversity: the value of mutual respect

NASA
> www.nasa.gov
>> Information on the space agency

Nobel Prize Internet Archive
> www.nobelprizes.com/nobel/
>> Nobel prize winners 1901-2000

Roget's Thesaurus
> www.thesaurus.com
>> Online version of the dictionary of synonyms

U.S. Department of Education
> www.ed.gov
>> How to get government educational loan

BROWSERS

A browser is the software you need to access the Web. It reads the HTML text and converts it to the Web page you want to see. Web browsers are usually part of the subscription to an Internet service provider such as Netscape Communicator or Microsoft Internet Explorer.

SEARCH ENGINES

Finding information on the Web can be easy and quick if you know how to use the search engines. Search engines are huge **indexes** of Web sites that serve as **research tools**. You must know what type of information you are looking for before selecting the engines that suit your needs. Remember, the most important step of your research strategy is to define the focus of your search. Once you have defined your topic, it is easy to choose the **appropriate keywords** and the **right tools** to search the Web. This is crucial to finding the information you need. The following are useful and well-known search engines.

Table 2.0.

Alta Vista: http://www.altavista.com—A powerful search engine with some unique features for advanced keyword searching. It supports Boolean operators (AND, OR, AND NOT including NEAR) which allows you to link keywords relevant to your topic. For example, if you are searching for information on student academic success you will type: student AND academic AND success. Everything is explained by clicking the "Help" link on any page.

Dogpile: http://www.dogpile.com—Searches several search engines simultaneously using your keywords. It automatically searches the indexes of Yahoo, Excite, Lycos, Alta Vista and several other databases. It supports Boolean operators (AND, OR, AND NOT including NEAR).

Excite: http://www.excite.com—Searches are performed by concept or keywords. It is devoted to finding information, regardless of what you need. It uses concept extraction to learn about relationships between words. A search on "social intelligence" also retrieves items containing the text "social behavior." Using Excite's Power Search improves results over simple keyword searches.

Google: http://www.google.com—An excellent search engine that employs text-matching techniques to find pages that are both important and relevant to a user's search. It presents information selected from news sources worldwide and searches pictures and images on the Web. It searches or browses through archives of Usenet news groups since 1995.

Hotbot: http://www.hotbot.com—A valuable tool for both beginning and advanced searchers. There are drop-down menus around the keyword form that helps you to refine your search from the beginning.

Infoseek: http://www.infoseek.com—A user-friendly search engine that supports Boolean operators AND, OR, NOT and phrase searching. A selection of options buttons placed under the keyword entry form allows you to distinguish the scope of your search.

Lycos: http://www.lycos.com—A basic Internet search engine. Searching from the home page leads you to almost anything you need, but the Advance Search option is a better choice to find what you need. It supports Boolean operators AND/OR.

WebCrawler: http://www.webcrawler.com—Effective at simple searches. It supports Boolean operators AND/OR/NOT to help you with advanced searches. Long search terms do not yield good results.

Yahoo! http://www.yahoo.com—Continues to hold its position as the number-one Internet search engine. It is also known as a Web directory. You can search a topic successfully with simple keywords or phrases, do not need to use Boolean operators. Due to the popular and scholarly topics of the Yahoo's pages, a search will yield smaller but more focused results.

ELECTRONIC MAIL OR E-MAIL

From the beginning of the Internet, the **electronic mail (e-mail)** has been the most widely used Internet service. Every system on the Internet supports some sort of mail service, regardless of the kind of computer used. In order to send and receive e-mail you need an **e-mail address** you can give people to get in touch with you. At the same time, you need the e-mail addresses of people with whom you want to communicate. Internet mail addresses are divided in two parts separated by @ (the *at* sign). The first part contains the username or identification number, and the second part, the **domain** name, contains the name of the user's Internet system or location. People with access to the Internet use the e-mail application to communicate with others from a distance in a quick and a somewhat informal way.

For additional information on the Internet you can visit the Web sites *Internet Tips* www.internet-tips.net/ and *Learn the Net* www.learn-thenet.com/english/index.html/.

How to Learn about the Library

For a start, find the reference and circulation desks, the different types of computer workstations, and the periodicals on display. Browse through the reference collection to find out what indexes the library subscribes in print and in electronic format. You should at least be familiar with the indexes you need. For example, if you plan to write a paper on a topic in education, the ERIC online database will help you identify full-text articles in that field. As a registered M-DC student you can access this research tool by visiting the North Campus Web page (www.mdcc.edu/north) and moving the mouse over the navigation bar where you will find a link to the library, or you can go directly to the library Web page at www.mdcc.edu/north/library and look for the link to the databases. To access the library databases all you need is your student ID. Remember that near the reference desk there are several computer workstations dedicated to Internet research only and that you can also use the computers housed in the Computer Courtyard located on the first floor of the library building.

Do not assume that librarians can understand libraries because they work full-time at it. Librarians will always understand libraries best because they manage library resources, but your job is to become an effective inquirer, not a librarian. In asking for help from the library staff, keep in mind that not everybody who works in the library is knowledgeable about all the library resources. You will find library technicians, library clerks, secretaries and even some of your peers assisting in the operation of the library. You have learned "why" one must follow a systematic approach to searching for information. Next you will learn more about "how to" use some of the research tools discussed in this chapter when you attend one of the library classes that the reference librarians teach. You will also learn a lot about your library through using the *Library Handbook for Students* available online on the North Campus Library Web page. The handbook discusses the materials available to you, how they are organized, how to access them and, if they are part of the circulating materials, how to check them out. In addition, you will learn more of the "how to" when you ask for help at the reference desk where there is **always** a librarian on duty.

Journal Questions/Activities

1. What is the best way to start your search for the publications you will use in your paper?

2. What can you do to improve the quality of the information you use in your papers?

3. Give at least two things you learned about searching for information in this chapter.

4. Did this chapter give you what you expected? What more do you need?

Summary

The literature of any field of study contains the system of communication you search for information. The literature includes all kinds of publications. There are three methods of locating publications on a subject of interest:

1. Looking up references cited in a basic book

2. Becoming familiar with publications

3. Using research tools.

A bibliography is a list of publications on a specific subject. A citation is a short description of a print or non-print publication and should be presented every time you use the writing of other authors.

To focus on your research, you must acquire background information on your topic so that you will be able to choose from the available publications. The best way to do this is to refer to a basic source such as an encyclopedia. The basic steps for research strategy are:

1. Define the focus of your search, and clarify unfamiliar terms

2. Break the topic into simple subtopics

3. Estimate the quantity of material needed

4. Specify the quality of material

5. Budget your time

6. Identify the information system you will search

7. Frame your question to match the language of the system.

The library is a logical organizational system with indexes to access the information. There are four major types of indexes or research tools in a library:

1. The online book catalog

2. Periodicals indexes (in print and electronic)

3. Abstracting services

4. Bibliographies

Finally, learn as much as possible about the library and the Internet. Knowing the location of sources related to your subject will help you in the process of finding the information you need.

Summary Exercise 3.2

True-False

Write an X in the appropriate line to indicate if the statement is true or false.

True   False

_____   _____   1.   Reading an article that deals with the focus of your interest is an excellent way of learning about a subject.

_____   _____   2.   A bibliography is a more or less comprehensive list of books and articles on a topic.

_____   _____   3.   You can omit the identification of publications used in your paper.

_____   _____   4.   The online library catalog enables you to search for books and articles owned by the college.

_____   _____   5.   Research strategy is the organization of the logical steps you must follow to solve a research problem.

Filling the blanks

Write in the space provided the appropriate word or words.

1.   You are committing _____ when you fail to acknowledge you used the information from a book or article to write your paper.

2.   A _____ is a short format used to identify the publication used in a paper.

3.   The first step in your research strategy is defining the _____ of your paper.

4.   The _____ is a worldwide communications system that links computers throughout the world.

5.   The _____ is just a portion of the Internet that enables you to link documents in different computers connected to the Internet.

Chapter Four

Motivation and Success

Carol Cooper, Ed.D.

Name _____   Date _____

Exercise 4.1

Success and Motivation Awareness Check

DIRECTIONS: Please place an "X" in the appropriate box.

	Yes	No	
1.	❐	❐	Success-oriented people attribute their success to skill and effort.
2.	❐	❐	Success-oriented people attribute their success to luck.
3.	❐	❐	Success-oriented people think positively.
4.	❐	❐	It is not always wise to plan too far ahead because many things turn out to be a matter of good or bad fortune anyhow.
5.	❐	❐	Daydreams can control your success in college.
6.	❐	❐	Our value system determines whether or not we will be successful college students.
7.	❐	❐	Our behaviors are shaped by society.
8.	❐	❐	Parents are probably the truest source of motivation.
9.	❐	❐	It is basically up to our teachers to motivate us to be successful in college.
10.	❐	❐	Teachers hold the key to my college success.

Introduction

Motivation is at the root of everything we do, think and feel. People who are motivated to achieve usually have positive attitudes toward the goal and everybody knows about it. They behave like they want to achieve. They can visualize the goal they want to achieve. Therefore, it's important that you understand the role motivation plays in your academic success. Studies have already shown us that student motivation is the best predictor of whether students will stay in school and get their degree. Have you not seen that some of the smartest students leave without their degree? It's clear they have the ability or credentials to succeed. According to Cope & Hannah, the reasons are due to reduced motivation, discouragement and disillusionment. Today we are bombarded with so many choices and wants. How do you narrow down and determine the choices you select. How do you know what you should go for? Do you know what you want? Once you make the choice, how do you know if you will stay with it?

Values play a role in determining what is important in life to you. As a result, values must be looked at in terms of their effect on motivation. Therefore, in this chapter, we will deal with (1) *the definition and stages of motivation*, (2) *the forces involved in motivating us*, (3) *who controls what we do, think and its effect on our lives*, (4) *the "Script for Success"* and (5) *characteristics of successful people*.

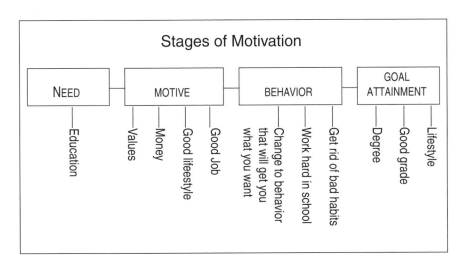

You dream about how successful you will be
and the kind of lifestyle you will live.
You set your sights on reaching your goal.
You motivate yourself.
You psyche yourself up,
and then you do it.

Characteristics of Motivated People

A strong desire to achieve
Clear needs & desires
Clear goals
Learn the necessary skills by
   preparing
They stay focused &
   disciplined
Persistent
Confident
Have a plan of action
Good communicator

Values, Beliefs

↓

Motive

↓

Motivation (Drive)

Motivation is that key or inner force that moves you through life. Motivation generates life energy and gives you direction. Your motive, which can be defined as an attribute of yourself, arouses, directs and integrates all your behavior. Everything you think, feel, and do comes from this inner force. Where does this inner force come from that makes you act? It usually develops through unmet needs based on your value system. Values are standards that determine how you live and behave. According to Stark, "Growing up involves making choices about which values you will adopt."[1] For most people, their value system comes from others such as parents and significant others whose approval they seek and/or want to emulate. However, in college, students may begin to look at value systems outside the family that may confuse them. This happens because of the vast number of people with differing life philosophies students come in contact with in college. It is natural to begin to question one's beliefs.

Let's take a quick look at your values. In the blanks below, list your top five values right now. In column A, just list them. In column B, prioritize them.

Column A	Column B
_____	5. _____
_____	6. _____
_____	7. _____
_____	8. _____
_____	9. _____

Discuss why you made the selections you did. Consider whether it is your value or someone else's value.

Regardless of which values you choose, your behavior will be shaped by the belief that something is missing in your life and you must seek to obtain it. How you obtain it is based on the strength of your belief that it is going to benefit you in a positive way. Is a degree in college worth the sacrifice? Will getting this degree help you acquire the kind of lifestyle you want in the future? True motivation can only come from within yourself. Parents and friends may encourage you in achieving a goal, but only you can actually make something happen. Self-goals are a powerful motivating force. You have to feel the need to act. If you want to learn and/or succeed, you have to intend to learn and/or succeed. In order to do this, you have to be motivated. If you want to be successful in life, you must have a goal and

---

[1].   M. Starke. *Survival Skills for College* (Englewood Cliffs: Prentice Hall, 1990) p. 222.

be motivated to go after it. The most important factors that will determine if you are going to succeed are having a self-motivated desire to get a good education, a positive attitude, the willingness to change and modify your behavior from one who does not work to one who does, take risks to learn, put in the required time for course preparation, and have a purpose along with good learning skills. Hopefully, over the years, you have learned to feel good about yourself and what you do because you will need that confidence if you are going to be successful. Students with negative self-concepts tend to lack confidence and, as a result, tend not to succeed. It's not too late to develop a positive self-concept. Take the free self-esteem test online at http://www.queendom.com/selfhelp. Motivation, then, explains your performance and/or why you behave as you do. In exercise 4.2 below, check off your goals and determine how your own preferences will affect what you do.

Exercise 4.2

BROAD LIFE GOALS (BASED ON ONE'S VALUE SYSTEM)

1. ☐ **Be Loving**

   To share and obtain companionship

2. ☐ **Be Dutiful**

   To dedicate myself to duty

3. ☐ **Be an Expert**

   To become an authority

4. ☐ **Be Independent**

   To have freedom in what I think and do

5. ☐ **Be a Leader**

   To become influential

6. ☐ **Be a Parent**

   To raise a family/have heirs

7. ☐ **Enjoy Life**

   To be happy and contented

8. ☐ **Have Power/Status**

   To have control of others

9. ☐ **Be Famous/Honored**

   To become well-known

10. ☐ **Be Secure**

    To have a stable position

11. ❐ **Serve Humanity**

To contribute to the satisfaction of others

12. ❐ **Be Wealthy**

To earn a great deal of money

13. ❐ **Others:**

_____

_____

_____

_____

_____

_____

_____

_____

How will your selections above shape your actions as a student? What must you do in order to achieve your goals?

_____

_____

_____

_____

_____

_____

_____

What are the barriers that will hinder you from reaching your goal?

_____

_____

_____

_____

How can you remove these barriers?

_____

_____

_____

_____

Exercise 4.3

Why Did You Decide to Come to College?

You could have elected to go into a full-time job and retire in that job twenty to thirty years later without having the expense of a college education. List FIVE REASONS for taking the college track as opposed to some other alternative.

1. _____
2. _____
3. _____
4. _____
5. _____

Now share your reasons and how they are going to motivate you in successfully getting through school. Did your list show that you were too dependent on other people in being here? If you are unable to tell, check with your professor or assigned small group if there is one. This phase is extremely important because it shapes the kind of character a student needs in order to be successful. Now look at the pros and cons of being in college if you have not already done so.

Exercise 4.4

College Yes, College No

—Attending College—

Pro	Con
_____	_____
_____	_____
_____	_____
_____	_____

In college, you are responsible for your learning. You must seek out the instructor for clarification if necessary. In order to be successful in the academic environment, you must know what it is you are striving for and why. If you have taken these steps, you have established a goal and can be said to be motivated. Whatever your reasons are for coming to college, one must always deal with reality in terms of what you must do now with your time and what you must put off for the future. You are responsible for your learning and it takes hard work. Are you able to do that? See how decisions are made in your life and who controls your behavior.

Exercise 4.5

The Importance of Control in Your Life

The following instrument has been designed to help you look at who influences your decisions and how you look at life. This will help you in determining if you are going to be able to achieve your goal. Do not look for right and wrong answers; simply respond honestly to each set of statements. Read each pair of statements and decide which statement seems more accurate. Now place a check mark before that statement.

1. ☐ a. A great deal that happens to me is probably a matter of chance.
   ☐ b. I am the master of my fate.

2. ☐ a. It is almost impossible to figure out how to please some people.
   ☐ b. Getting along with people is a skill that must be practiced.

3. ☐ a. People like me can change the course of world affairs if they make themselves heard.
   ☐ b. It is only wishful thinking to believe that one can really influence what happens in society at large.

4. ☐ a. Most people get the respect they deserve in this world.
   ☐ b. An individual's capabilities often pass unrecognized no matter how hard he or she tries.

5. ☐ a. The idea that teachers are unfair to students is nonsense.
   ☐ b. Most students do not realize the extent to which grades are influenced by accidental happenings.

6. ☐ a. If one doesn't get the right opportunities, one cannot become a good leader.
   ☐ b. Capable people who fail to become leaders have not taken advantage of their opportunities.

7. ☐ a. Sometimes I feel I have little to do with the grades I get.
   ☐ b. In my case, the grades I make are the results of my own efforts; luck has little or nothing to do with it.

8. ☐ a. Heredity plays the major role in determining one's personality.
   ☐ b. It is one's experience in life which determines what one is like.

9. ☐ a. It is silly to think that one can really change another person's basic attitudes.
   ☐ b. When I am right, I can convince others.

10. □ a. In my experience, I have noticed that there is usually a direct connection between how hard I study and the grades I get.

□ b. Many times the reactions of teachers seem haphazard to me.

11. □ a. The average citizen can have an influence in government decisions.

□ b. The world is run by the few people in power, and there is not much the little guy can do about it.

12. □ a. When I make plans, I am almost certain that I can make them work.

□ b. It is not always wise to plan too far ahead because many things turn out to be a matter of good or bad fortune.

13. □ a. There are certain people who are just no good.

□ b. There is some good in everybody.

14. □ a. In my case, getting what I want has little or nothing to do with luck.

□ b. Many times we might just as well decide what to do by flipping a coin.

15. □ a. Promotions are earned through hard work and persistence.

□ b. Making a lot of money is largely a matter of getting the right breaks.

16. □ a. Marriage is largely a gamble.

□ b. The number of divorces indicates that more and more people are not trying to make their marriages work.

17. □ a. In our society a man's future earning power is dependent upon his ability.

□ b. Getting promoted is really a matter of being a little luckier than the next guy.

18. □ a. Most people don't realize the extent to which they are controlled by accidental happenings.

□ b. There is really no such thing as luck.

19. □ a. I have a little influence over the way other people behave.

□ b. If one knows how to deal with people, he/she knows they are really quite easily led.

20. □ a. In the long run, the bad things that happen to us are balanced by the good ones.

□ b. Most misfortunes are the result of lack of ability, ignorance, laziness, or all three.

21. ☐ a. With enough effort, we can wipe out political corruption.

    ☐ b. It is difficult for people to have much control over the things politicians do in office.

22. ☐ a. A good leader expects people to decide for themselves what they should do.

    ☐ b. A good leader makes it clear to everybody what his/her job is.

23. ☐ a. People are lonely because they don't try to be friendly.

    ☐ b. There's not much use in trying to please people. If they like you, they like you.

24. ☐ a. There is too much emphasis on athletics in high school.

    ☐ b. Team sports are an excellent way to build character.

25. ☐ a. What happens to me is my own doing.

    ☐ b. Sometimes I feel I don't have enough control over the direction my life is taking.[1]

[1.] Walker, Velma & Lynn Brokaw. *Becoming Aware, A* Human Relations Handbook, 4th ed. Iowa: Kendall/Hunt Publishing Company, 1981.

SCORING KEY

DIRECTIONS: If you selected the responses below, give yourself one (1) point. For example, if for Item No. 1 you selected the "a" statement, give yourself one (1) point.

Item#	Response	Item#	Response	Item#	Response
1	A	10	B	19	A
2	A	11	B	20	A
3	B	12	B	21	B
4	B	13	A	22	B
5	B	14	B	23	B
6	A	15	B	24	A
7	A	16	A	25	B
8	A	17	B		
9	A	18	A		

EXPLANATION AND DISCUSSION

Julian Rotter, a psychologist, states that people who believe that they are the master of their own fate and control what happens to them are said to have an internal locus of control. People who believe that their fate is in the hands of others such as the teacher, government and luck are said to have an external locus of control.

THE FAR SIDE® BY GARY LARSON

"Mr. Osborne, may I be excused?
My brain is full."

What is the significance if you are labeled one or the other? Students who believe in themselves and have a high internal locus of control make things happen. They take responsibility for controlling what happens in their lives. On the other hand, students with high external locus of control seldom, if ever, initiate any action. They allow others to make the decision. They follow the crowd and appear to be easily influenced. In other words, they do not control what happens to them.

Internal locus of control

External locus of control

INTERNAL? EXTERNAL?

Your total number of points indicates whether you are internal or external. The higher the score above fifteen (15), the more you perceive control as being external. The lower the score below fifteen (15), the more you perceive control as internal. Where did you fall?

Identify your locus of control in the blank space.

_____; If you want to be successful in college and life in general, what behaviors can you change and/or improve right now? Have you heard the statement, "we are what we think?" So what is it that you want to think about so that you become what you want to be?

_____

_____

_____

_____

Misconceptions about Success

Before we look at how to be successful, let's look at some misconceptions about success.

- *You have to be in the right place at the right time.* What if someone did present you with a great job? Would you have the skills required to maintain the status quo.

- *It takes luck to be successful.* For some people luck may be a part of it, but hard work is what you need more—smart hard work.

- *Background and culture determines whether or not you will be successful.* No. Anybody can be successful.

- *Successful people don't make mistakes.* Yes, they do make mistakes. However, they learn not to repeat them. These mistakes become challenges to stepping stones for them.

- *Successful people just seem to know how to do things.* No. That's not true either. We don't live in isolation and will need to learn and share with others as we climb the ladder of success. In the sharing process is when we learn how to do things.

Script for Success

For sure, you must take control of your life and begin to make things happen. You already know that in the academic environment, you have to be motivated and put the necessary time into classroom activity and study activity if you want to be successful. You have already made the decision to be here. Why not begin to think positively about your educational experience and be the best that you can be in each and every class? Self-control means you set your priorities in life based on your goals and you set a time-frame in which you plan to achieve them.

STEPS TO SUCCESS

Day dreaming

1. Think success in every class. David McClelland, a Harvard psychologist, conducted research in the 1960's and proved that he could predict whether a student would be successful based on his/her daydreaming (imagination). He even proved that individuals could be taught to be successful based on the way they thought about tasks and achieving goals. How can you begin to think successfully? Follow these steps:

Steps to successful thinking

First, evaluate the task and set realistic and obtainable goals. Always work within the given time-frame. Make a step-by-step plan in trying to achieve the goal. Don't bite off too much at one time. If you do, you will fail. You must not only be successful but also see yourself being successful and striving to do better.

Second, complete easier tasks first. Success breeds success.

Third, never allow an obstacle to get in your way. This obstacle may be a friend, social event, time perception or a lack of understanding of the assigned task. You can always work yourself through, around or under obstacles. Can't think of how to do this? Seek other resources (different points of view) and determine a plan of action. Believe that there is a solution to every problem. You must believe that you are in control and can be successful at whatever you try.

Fourth, always reward yourself for a task well done.

2. Look at school as a positive experience and plan to do well in every class.

3. If you can find a job tutoring other students on campus, do it. When you must teach someone else, it means you must learn the material well first.

4. Learn to think critically; then put the skill into practice. Think objectively for yourself.

Exercise 4.6

Script for Success

The list above is incomplete and only you can complete it. Think of some successful people you know and/or with whom you are familiar. Make a list of their successful behaviors. Which would you begin to use in your life? Use the space below to complete this exercise. In the first column, write the name; in the second column, identify their career; and in the third, list characteristics of that person's successful behavior.

Name   Career   Characteristics

_____

_____

_____

_____

_____

_____

_____

Characteristics of Successful People

1. Success takes hard work and a person who is willing to do it.   **Hard worker**
2. When problems arise, they are faced. The problems are looked at as   **Problem solver**
    opportunities to excel.
3. Successful people do more than what is required. They go one step   **Overachiever**
    beyond.
4. They persevere and do not quit. Yes, they may encounter setbacks,   **Persistent**
    but these do not stop them.
5. They believe in themselves and that they will succeed. They tune out   **Positive thinker**
    words such as "no" and "can't."
6. They believe they are responsible for putting in the required time and **Responsible**
    effort required to achieve their goal.
7. They always assess their environment before making decisions.   **Realistic**

Adapted from *Up from Underachievement* by Felton & Briggs.

Vision-directed   8.   They write down their vision and always keep it in front of them. They make use of any opportunity that allows them to move one step closer to making it happen. They give their vision high priority in their lives.

Honest   9.   They are honest about themselves and with the people with whom they work. They know when to seek help.

Lifelong learner   10.   They always strive to learn and grow. They know that there is always something to be learned and that they don't know everything.

Helper   11.   They give of themselves to others. They lend a helping hand to others.

Equipment

Figure it out for yourself my lad
You've all that the greatest of men have had,
Two arms, two hands, two legs, two eyes,
And a brain to use if you would be wise,
With this equipment they all began,
So start from the top and say, "I Can."
Courage must come from the soul within,
The man must furnish the will to win,
So figure it out for yourself my lad,
You were born with all that the great have had.
With your equipment they all began,
Get hold of yourself and say: "I Can."

Anonymous

 Journal Questions/Activities

1.   Log on to the Internet. Go to website http://queendom.com/selfhelp and click into the free self-esteem test. Take the inventory and share your findings. If it recommends that you need to make changes, share what you plan to do and how you plan to go about it. Ask your instructor for assistance. If that site is not available, ask your instructor for assistance in coming up with another alternative.

2.   Using the stages of motivation as a guide, share your a) needs, b) motives, c) behaviors and d) expected goal attainments.

Summary

As you can see from this chapter, your motivation reflects how you understand yourself. This understanding then causes you to respond to life in a particular manner. Our values/belief system determines our goals. These goals can be looked at as being motives that arouse and/or direct our behavior. The need to achieve these goals is the force that is going to motivate us.

In this chapter, we have covered the role of values and how they influence decisions in our lives. We have stressed that the reason you are in college and your locus of control will have a determining effect on whether you will be successful in school and life in general. The script for success and the characteristics of success-oriented people have been included to help you look at where you are and to see if you need to make any changes if you want to be successful.

Name _____   Date _____

Summary Exercise 4.7

Review Chapter Test

1. Define motivation and discuss how it can have a negative or positive effect on obtaining one's educational goal(s).

2. Compare and contrast the following:

    Internal Locus of Control

    External Locus of Control

3. In relationship to you or other college students, what was the significance of David McClelland's study?

4. Using the box below, identify two obstacles you may encounter as a college student and what you can do to overcome each of them. Be specific in your plans.

Obstacles	Plan of Action

References

A. Merriam-Webster. *Webster's Tenth New Collegiate Dictionary.* Springfield, Massachusetts: Merriam-Webster, Inc., 1988. http://www.m-w.com.

Anderson, E.C. "Forces Influencing Student Persistence and Achievement." In L. Noel, R.S. Levitz, & D. Soluri *Increasing Student Retention.* San Francisco: Jossey-Bass 1985.

Beilke, Ines Torres. *Career Motivation and Self-Concept.* Dubuque, Iowa: Kendall/Hunt, 1986.

Chapman, Elwood N. *College Survival. A Do It Yourself Guide,* 2nd ed., Chicago: Science Research Associates, Inc., 1981.

Gardner, John & Jewler, A. Jerome. *College Is Only the Beginning. The Freshman Year Experience Series,* 2nd ed., Belmont, Calif.: Wadsworth Publishing Co., 1989.

Goethals, George R. & Worchel, Stephen. *Adjustment and Human Relations.* New York: Alfred A. Knopf, Inc., 1981.

Landsberger, Joe. *Study Guide & Strategies,* University of St. Thomas, St. Paul, Minnesota, 2003. http://www.studygs.net/motivation.htm.

Nikelly, Arthur G. *Achieving Competence and Fulfillment.* Monterey, Calif.: Brooks Cole Publishing Company, (1977) pp. 119-120.

Rathus, Spencer A., *Psychology.* New York: Holt, Rinehart and Winston, 1981.

Waitley, Denis. *Psychology of Success: Finding Meaning in Work and Life,* 4th edition. New York: McGraw-Hill, 2004.

Walker, Velma & Brokaw, Lynn. *Becoming Aware, A Human Relations Handbook,* 4th ed. Iowa: Kendall/Hunt Publishing Company, 1981.

Walter, Tim & Siebert, Al. *Student Success: How To Succeed In College and Still Have Time For Friends,* 5th ed., Fort Worth, Texas: Holt, Rinehart and Winston, Inc., 1990.

http://www.queendom.com/selfhelp. 1996-2004.

Chapter Five

Time Management

Carol Cooper, Ed. D

Name _____   Date _____

Exercise 5.1

Time Management Awareness Check

DIRECTIONS: Place an "X" in the appropriate box and check your responses with the answer key.

	Yes	No	
1.	❏	❏	Do you make a calendar and plan your coursework for the entire semester?
2.	❏	❏	Do you follow a weekly time schedule?
3.	❏	❏	Do you use a daily calendar or notebook to keep you aware of high priority items to do now?
4.	❏	❏	When you take a course, do you know the criteria for assigning grades?
5.	❏	❏	Are there times when you have difficulty starting to study?
6.	❏	❏	Is it sometimes necessary for you to cram for a test?
7.	❏	❏	Do you have trouble getting to class on time?
8.	❏	❏	Do you have difficulty concentrating on an assignment?
9.	❏	❏	Is it sometimes hard for you to finish term papers, reports and projects on time?
10.	❏	❏	Are you easily distracted when studying?
11.	❏	❏	Do you become bored with the subject when studying?
12.	❏	❏	Do you reward yourself when you have studied effectively?
13.	❏	❏	Do you schedule large blocks of time for study?
14.	❏	❏	Do you have a comfortable place where you study regularly?
15.	❏	❏	Do you put off starting on a big assignment because you think it is too hard?
16.	❏	❏	Do you use short periods of free time for studying effectively?
17.	❏	❏	Do you plan time for rest periods when studying?
18.	❏	❏	Do you plan time for recreation and relaxation?
19.	❏	❏	Do you panic or become anxious in test situations?
20.	❏	❏	Do you study your hardest subjects when you are more alert?
21.	❏	❏	When you have a high priority assignment, do you work at smaller, routine jobs instead of the important one?

Model Answer Key

1. Yes	8. No	15. No
2. Yes	9. No	16. Yes
3. Yes	10. No	17. Yes
4. Yes	11. No	18. Yes
5. No	12. Yes	19. No
6. No	13. Yes	20. Yes
7. No	14. Yes	21. No

Author Unknown

Few people achieve all these ideal answers. As you go through this chapter, examine your "wrong" responses to see if you can find a better system than you have at present.

Introduction

In this chapter, we will discuss how to set goals, how to make efficient time schedules and how to break the habit of procrastinating. The major objective in this chapter is to help you integrate your academic and non-academic life into a realistic and workable schedule so that you can achieve your goals. We will offer you many tips on how to plan your time around social/personal obligations, school and work.

How you use your time and what motivates your use of your time are probably the most important factors which will determine if you will be successful. One of the major reasons first-year students fail is their inability to effectively manage their time. Get smart and take control of your time and don't let it control you. According to Alan Lakein, "Time is life. It is irreversible and irreplaceable. To waste time is to waste your life, but to control time is to master your life and make the most of it."[1]

If you already know the SECRET to time management, you probably scored very well on the self-check. If you do not, you probably fell into a time management TRAP. You will also learn that in order for any skill to be developed, you must practice it, thereby learning the secret.

With all the daily pressures, it is very easy to lose track of time. Some of these pressures come about because of today's value system which presents another way of looking at time. In a multicultural society, time is usually perceived based on one's cultural heritage. A person's use of time is based on his/her value system. Its use is learned behavior. The whole issue of time is important depending on where a person works and plays and whose orientation governs the process. For example according to Stewart, Americans are future-oriented in terms of work and action. They believe that hard work now will bring about success in the future. No matter what your temporal orientation of time is, you must adapt it to the society you want to work and play in, if you want to be successful. This is not an easy task since you are talking about modifying and/or changing values/behaviors. Why? Simply put, we allow time to control us. Let's look at a major time-management problem.

Procrastination

Procrastination is the act of putting off tasks that you need to do now not sometime in the future. Sometimes, as a result of putting the tasks off, they never get done.

---

[1]   W. Herlin and C. Mayfield. *Successful Study Skills* (Dubuque: Kendall/Hunt Publishing Company) p. 9. 1981.

People who procrastinate tend to:

- set unrealistic goals

- see the task as overwhelming

- fear that they will fail

- have low level of frustration tolerance

- overestimate or underestimate time needed to complete a task

- sometimes doubt their own capabilities

- lack the motivation to get started

The following steps are recommended to break the cycle of procrastination. Try them and see what happens.

1. Define your goal. It will give you a sense of direction. It will help you determine what you need to do.

2. Determine what your needs are.

3. Assess the task at hand. If it's very large, break it down into smaller units of work. Recognize that some assignments will be harder than others.

4. Organize by setting up a plan of action to get the work done. Each task must have a target date for completing it and a specific time of day. DO NOT DEVIATE. Try completing a portion of the work every day.

5. Get to work. Set your clock. Try it out. Breaking a habit takes about three weeks of consistent work. Reward yourself when you complete the assignment.

6. Reevaluate if your plan of action does not work. BE PERSISTENT. DO NOT QUIT.

Goal-Setting

Manage your time. Plan your life. *In case you have not caught on yet, you are talking about managing yourself. You are talking about getting rid of habits that tend to interfere with what you need to do.* You need to set goals and priorities around your commitment to obtain a college education and a profession. Managing your time will help you successfully achieve these commitments. If you are having problems on the job, evaluate the situation, set goals and give yourself a specified period of time to complete them.

Time management calls for planning. Always ask yourself, what should I be doing right now? You cannot plan without making decisions about what you want to do and where you want to go. *Effective time management requires you to be a decision-maker.* When you make the decisions about what you want and where you want to go, you have established GOALS. *Goals are important because they allow you to put your values into action. Goals are no more than your aims in life and they give you purpose and direction in which to focus your energies.*

To develop a goal is the first step in time management. *The next step is to write your goal down in explicit and concrete terms. Until it is written, it is just an idea. A goal should have the following characteristics:*

Reachable—Set it up in small increments so you don't bite off more than you can chew at one time. If you *do, you will only become frustrated.* Set only moderate goals you know you can reach.

First step in time management

Characteristics of goals

Realistic—*Know your limitations and capabilities. Don't ask more of yourself than you know you are capable of doing. Set a realistic time-frame to achieve it.*

Measurable—*State your goal in such a way that you as well as others will know when it has been achieved.* Be very specific and concrete about what you want to do.

For each goal, always write out a step-by-step plan as to what you are going to do to achieve it. Be concrete and specific. Each goal should always have a time-frame. Without a time frame, some people would put-off and put-off and never carry out the plan.

Goals are described as short-term and long-term. Long term goals usually take a while to accomplish while short term goals help you achieve the long-term goal. They are activities you carry out on a daily, weekly or monthly schedule.

Second step in time management

Now that you have your goal, you must develop a "TO DO LIST" which may be daily, weekly, or monthly. Then PRIORITIZE. As a student, you should develop a "daily to do list." Look at an example of how goals and priorities can be utilized.

Long-Term Goal

I want to get my associate in arts degree in business from Miami-Dade College in 200_.

Short-Term Goal

I want to pass English (ENC1101) and Social Environment (ISS1120) with a grade of "C" or above this semester.

Daily To-Do List
August 31, 20__.

Prioritize your daily "to-do" list. Ask yourself the following question. What must I get done first? If I don't get it done, what will be the consequences.

Rank
Ordering
4   Visit with Mark and Gary
__ See "Star Trek" movie
3   Practice essay writing
__   Watch television
1   Go to work 8:00 a.m. to 12:00 noon
2   Study chapters 1 & 2, ISS1120

According to the example above, the number one priority on August 31, 20__ is to go to work. The second priority is to study ISS1120, and the third is to practice essay writing. This is good prioritizing since this student's short-term goal is to pass English and social environment.

Remember, you must set your goals and then begin to organize your time by using "to-do" lists which you prioritize. Most students tend to function better if they divide their goal-setting into three levels:

DAILY—WEEKLY—SEMESTER

Before you start each week, jot down the tasks you need to get done. Prioritize them and place them in your weekly schedule. As you get them done circle them in blue. This means you have accomplished the goal for the week.

Ideally, buying an organizer and taking it with you everywhere you go is the thing to do. Successful students plan to complete tasks ahead of schedule.

Goal-setting should be established at the beginning of each term. This includes analysis of information on course selection, what grades you want to receive in each course, how you plan to go about it and the evaluation of your first class sessions. Put all of this information together before setting time schedules. Remember, a goal is no good unless it is explicit and concrete. For example, to say you want to do well would be too general and vague. "I plan to make an 'A' in Psychology" is better.

Exercise 5.2

SETTING GOALS

DIRECTIONS: Write out at least two of your long-term goals in the first two rows in columns A and B. At least one of them must pertain to your education or future profession. Then list at least four short-term goals under each long-term goal.

Column A

Long-Term Goal #1

_____

_____

_____

Short-Term Goals

_____

_____

_____

Column B

Long-Term Goal #2

_____

_____

_____

Short-Term Goals

_____

_____

_____

Planning Your Time?

If you have not purchased your academic calendar planner for the year, you should rush right out and do it now. It should be large enough so you can write in your assignments. The ideal one would also have space for your daily "to-do" list. However, do not worry about that space since all you have to do is buy a pack of 3"x5" cards and on a nightly basis before going to bed, write out all the tasks you are planning for the next day. Don't forget to prioritize. Before writing in your planner, use these tips when planning your time.

Tip 1:   Balance your time.

Balance your work, travel, sleep, domestic chores, class, study, personal, and recreational times. All work and no play truly do make Jack a dull boy. You will no doubt become frustrated if you spend too much time in some areas and your goals are not being met in other areas.

Tip 2:   Plan for the semester.

Plan your school work for an entire semester based on the school's calendar. This refers to due dates for papers, hours for study time, dates for major exams, etc.

Tip 3:   Set goals for each study session.

Always set a goal for each study session and be definite in your schedule about what you plan to study. Be definite about what you plan to do in each session.

Tip 4:   Know how long your study session should be.

Study in short sessions. The idea is to avoid marathon sessions where you remember little of what you tried to learn. Plan to take at least a ten-minute break during every hour. During the break, try exercising or doing something else that will refresh your mind/body. This will help you maintain your concentration.

Tip 5:   Study in the right place.

Select a quiet and not-too-comfortable place to study. It should allow you to concentrate on the task at hand. Make sure it has proper lighting, ventilation and a comfortable temperature.

Tip 6:   Let others in on your plans.

Let your friends and significant others know your study schedule so they will not disturb you. You are the person who must control someone else's use of your time. If you have children, plan activities for them while you study.

Tip 7:   Plan to see your professor.

Make sure you understand your notes and/or assignment before trying to study. If you are not clear on what you have to study, make an appointment to visit your professor for clarification. Don't flounder. Record your appointment to see the professor if it is necessary in your appointment book.

Tip 8:   Know how much time you should study for each class.

Plan two hours of study for every hour you are in class.

Tip 9:   Prioritize subjects to study.

Study the subject you like least first since it will no doubt require more of your time and energy. Once you have completed this task, reward yourself by doing something you like.

Tip 10: Know when to review notes.

Review, study and/or rewrite your lecture notes within 24 hours to help with memory and effective note-taking. Forty-eight hours should be the maximum amount of time you allow to lapse before reviewing.

Tip 11: Know purpose of studying.

Study to pass tests. When reading and/or reviewing notes, always practice asking and answering questions. Study as though at the end of your study session, you will be required to pass a test.

Tip 12: Know how to begin reading text material.

Survey required chapters (material) before you begin to read.

Tip 13: Develop questions you should ask as you study.

Ask questions about what you must learn during the study period. Turn all headings into questions and then answer them.

Tip 14: Always read with a purpose.

Read the assigned chapters and/or material. Look for answers to questions posed in order to complete the assignment. Read with the purpose of finding the answers.

Tip 15: Memorize.

Go over the content which you want to remember. If necessary, orally recite and make notes to help you remember.

Tip 16: Review.

Review the material and ask questions.

Tip 17: Decide when you should study.

Determine your best time of day and schedule your study time then. In addition, a brief review before class and immediately after class is strongly suggested.

Tip 18: Use your time productively.

Don't waste time. If you are waiting, use that time for review. If you have recorded your notes and you are driving, listen to the tape. If you are riding with someone else, read your notes.

The third step in time management

After you have prioritized and come up with a plan for accomplishing your goals is to simply do it. This step will not be successful if you have not taken into consideration all of your responsibilities and carefully thought out the question,

What is the best use of my time right now if I want to accomplish my goals for the future, yet live the best I can right now?

Taking Control of Your Time

Now that you have looked at goal-setting and establishing priorities, you are ready for the next phase in understanding how you use time-taking control. The first step is to see how you actually use your time. Once you truly know what you are doing with your time, you should plan a weekly schedule for the semester using the time management tips in this chapter. Remember there are only 168 hours in a week.

Exercise 5.3

MONITORING YOUR TIME

1. Complete a daily log for a full week indicating how your time and energy were spent. Follow these directions:

   A. List all fixed obligations—classes, meetings, work hours, meals, travel time to and from commitments, family obligations, etc.—as you complete them.

   B. Consider and indicate time spent on class review if you prepared for class.

   C. Don't forget to indicate any time spent on health essentials such as recreation, sleep and exercise.

   D. Be sure 24 hours per day are accounted for in your log.

   E. Use the sample time chart on the following page.

2. Summarize the time you spent each day on various activities according to the summary log sheet provided.

Recommended website for goals, day and weekly planners: `http://www.iss.stthomas.edu/studyguides/schedule/`

SAMPLE PERSONAL LOG OF TIME USAGE

Name_____

Student Number_____ Date_____

Fill in each time block **after you have completed the activity.** Use the following categories of activities

Class	Travel	Exercise
Work	Eating	Personal (getting dressed, bathing,
Housework	Sleep	brushing teeth, etc.)
Meeting	Recreation	Other (explain)

SAMPLE

HOURS	MONDAY	TUESDAY	WEDNESDAY	THURSDAY	FRIDAY	SATURDAY	SUNDAY
7:00 A.M.	breakfast	dress	breakfast	dress	breakfast	sleep	sleep
7:30	travel	breakfast	travel	breakfast	travel	dress	" "
8:00	history class	travel	history class	travel	history class	breakfast	dress
8:30	" "	study speech	" "	study speech	" "	travel	breakfast
9:00	study history	" "	study history	" "	study history	work	family
9:30	" "	speech class	" "	speech class	" "	" "	commitment
10:00	English class	" "	English class	" "	English class	" "	" "
10:30	" "	" "	" "	" "	" "	" "	travel
11:00	study English	biology class	study English	biology class	study English	" "	church
11:30	" "	" "	" "	" "	" "	" "	" "
12:00 Noon	lunch	" "	lunch	" "	lunch	lunch	travel
12:30 P.M.	" "	study biology	" "	study biology	" "	travel	recreation
1:00	travel	" "	travel	" "	travel	errands	" "
1:30	errands	lunch	errands	lunch	errands	" "	" "
2:00	work	travel	work	travel	work	study history	" "
2:30	" "	work	" "	work	" "	" "	" "
3:00	" "	" "	" "	" "	" "	study English	" "
3:30	" "	" "	" "	" "	" "	" "	travel
4:00	" "	" "	" "	" "	" "	study biology	family
4:30	" "	" "	" "	" "	" "	" "	commitment

SAMPLE PERSONAL LOG OF TIME USAGE

SAMPLE

HOURS	MONDAY	TUESDAY	WEDNESDAY	THURSDAY	FRIDAY	SATURDAY	SUNDAY
5:00 P.M.	travel	work	travel	work	travel	relax	family
5:30	relax	travel	relax	travel	relax	" "	commitment
6:00	dinner	dinner	dinner	dinner	dinner	dinner	" "
6:30	" "	" "	" "	" "	" "	" "	dinner
7:00	study history	study speech	study history	study speech	recreation	recreation	" "
7:30	" "	" "	" "	" "	" "	" "	study or work
8:00	study English	study biology	study English	study biology	" "	" "	on special
8:30	" "	" "	" "	" "	" "	" "	class assign-
9:00	recreation	recreation	recreation	recreation	" "	" "	ments
9:30	" "	" "	" "	" "	" "	" "	" "
10:00	" "	" "	" "	" "	" "	" "	" "
10:30							
11:00							
11:30							
12:00 MIDN.							
12:30 A.M.							
1:00							
1:30							
2:00							
2:30							
3:00							
3:30							
4:00							
4:30							
5:00							
5:30							
6:00							
6:30							

Monitoring Your Time—Exercise 5.3

Name_____

Student Number_____ Date_____

Fill in each time block **after you have completed the activity.** Use the following categories of activities

Class	Travel	Exercise
Work	Eating	Personal (getting dressed, bathing,
Housework	Sleep	brushing teeth, etc.)
Meeting	Recreation	Other (explain)

HOURS	MONDAY	TUESDAY	WEDNESDAY	THURSDAY	FRIDAY	SATURDAY	SUNDAY
7:00 A.M.							
7:30							
8:00							
8:30							
9:00							
9:30							
10:00							
10:30							
11:00							
11:30							
12:00 Noon							
12:30 P.M.							
1:00							
1:30							
2:00							
2:30							
3:00							
3:30							
4:00							
4:30							

HOURS	MONDAY	TUESDAY	WEDNESDAY	THURSDAY	FRIDAY	SATURDAY	SUNDAY
5:00 P.M.							
5:30							
6:00							
6:30							
7:00							
7:30							
8:00							
8:30							
9:00							
9:30							
10:00							
10:30							
11:00							
11:30							
12:00 MIDN.							
12:30 A.M.							
1:00							
1:30							
2:00							
2:30							
3:00							
3:30							
4:00							
4:30							
5:00							
5:30							
6:00							
6:30							

Exercise 5.3
Monitor Summary

Name _____   Date _____   (Indicate Hours Spent in Each Area)

Student # _____

ACTIVITY	MONDAY	TUESDAY	WEDNESDAY	THURSDAY	FRIDAY	SATURDAY	SUNDAY	TOTAL
CLASSES								
WORK								
STUDYING								
TRAVEL								
RECREATION								
EXERCISE								
EATING								
FAMILY								
PERSONAL								
SLEEP								
OTHER (EXPLAIN)								
	24	24	24	24	24	24	24	168

Exercise 5.4

Analyzing Your Time

According to your personal summary log, how did you spend your time?

Identify the eight areas where you spent the greatest amounts of time. Explain why. Were any of them time wasters? Time wasters are behaviors getting in the way of you completing your desired goals. Some examples are: daydreaming, too many telephone calls, watching TV, partying, too much time on relationships, inability to say no, and sleeping too much.

Look at what effects these time wasters are having on your life.

_____

_____

_____

_____

What areas should you cut back on if you are going to accomplish your goals?

_____

_____

_____

How many hours will you reduce each one by?

_____

_____

In what areas should you increase the amount of time you spend if you are going to accomplish your goals?

_____

_____

_____

How many hours will you increase each one by?

_____

_____

Are there any other factors you should consider before actually planning and taking control of your time? If so, list them below and discuss them.

_____

_____

Exercise 5.5

PLANNING YOUR TIME

This is a very important exercise because if you really follow the time management tips and are highly motivated, there is no way you can fail. The purpose of a time schedule is to give you a framework for bringing order and discipline into your life, not to make you into a robot. It will give you time to do the things you need to do. It can break the pattern of procrastinating and cut down on worrying time. **Assume that you are basically planning how the rest of your weeks will look during the semester** with the exception of a few changes for unplanned events.

1. Complete a daily log for a full week indicating how you **PLAN** to spend your time.

    A. List all fixed obligations—classes, work, domestic chores, travel to and from commitments, sleep, and meals—you know you must complete.

    B. Don't forget to allow time for yourself (personal time), recreation, religion, and exercise.

    C. Based on class times, indicate when you plan to study and review school work. Don't forget to consider the amount of time you need for every hour in class and when you should review for information to move from short-term to long-term memory.

    D. Go back to Exercise 5.4 and use the information from your analysis.

    E. Be sure to plan for 24 hours a day:

2. Now summarize the time you plan to spend each day on the various activities according to the planning summary log sheet.

Exercise 5.5
Planning Summary

Name _____   Date _____

Student # _____   (Indicate Hours Spent in Each Area)

ACTIVITY	MONDAY	TUESDAY	WEDNESDAY	THURSDAY	FRIDAY	SATURDAY	SUNDAY	TOTAL
CLASSES								
WORK								
STUDYING								
TRAVEL								
RECREATION								
EXERCISE								
EATING								
FAMILY								
PERSONAL								
SLEEP								
OTHER (EXPLAIN)								
	24	24	24	24	24	24	24	168

? Journal Questions/Activities

Complete the following questions and statements.

1. Why do I continue to do the same things over and over when the behavior(s) clearly is not getting me where I want to go?

2. Is there some behavior in my life I need to change? Explain your answer.

Summary

Managing time, procrastinating and setting effective goals are major problems for students. This chapter has offered six steps to break the habit of procrastinating. The first step is to set reachable goals. All goals must be reasonable, realistic and measurable and controllable. If you have many tasks that you must complete, then you have to make a "to do" list and then prioritize your activities. A number of tips are given to help you make the most of your time. The second step is to determine your needs. The third is to assess the task or situation. The fourth step is to set up a plan of action with a specific time frame for achieving the goal. This is the time to map out in a schedule how much time you are going to use each day and when you are going to so this task. The fifth step is to get to work and try out the plan. The final and sixth step is to reevaluate and see if the plan is working.

Name _____   Date _____

Summary Exercise 5.6

1. What is the SECRET to time management?

    _____

    _____

2. Identify what you think are the three most important goals in your life and map out how you plan to go about achieving them.

    _____

    _____

    _____

3. Why is it so important to use a daily "to do" list?

    _____

    _____

4. Develop a plan of action showing how you would assist students to stop procrastinating. Be ready to share and discuss this plan with the class.

    _____

    _____

    _____

    _____

    _____

    _____

    _____

    _____

    _____

    _____

References

A. Merriam-Webster, *Webster's Tenth New Collegiate Dictionary*. Springfield, Massachusetts: Merriam-Webster, Inc., 1998.

Achievement Center. Time Management Principles, University of Minnesota, Duluth, 2002. `http://www.d.umn.edu/student/loon/acad/strat/time_man_princ.html`.

Beck, John A. and Clason, Marmy A. *On the Edge of Success*. Belmont, Ca.: Thomson Wadsworth, 2003.

Ellis, David. *Becoming a Master Student,* 10th edition. New York: College Survival, Inc., Houghton Mifflin Co. 1998.

Herlin, Wayne and Mayfield, Craig. *Successful Study Skills*. Dubuque, Iowa: Kendall/Hunt Publishing Co., 1981.

Landsberger, Joe. The Study Guides and Strategies websites, University of St. Thomas, St. Paul, Minnesota. 2003. `http://www.iss.stthomas.edu/studyguides/schedule/`.

Orientation and Study Skills Staff and Faculty, Miami-Dade Community College, Miami, Fla.

Personal Time Management and Goal Setting Guide. `http://www.time-management-guide.com/`.

Siebert, Al and Walter, Tim. *Student Success: How to Succeed in College and Still Have Time for Friends.* 5th edition. New York: Holt, Rinehart and Winston, 1990.

Starke, Mary C. *Survival Skills for College*. Englewood Cliffs, New Jersey: Prentice Hall, 1990.

Chapter Six

Listening and Note-Taking

Sheryl M. Hartman, Ph.D.

Exercise 6.1

Listening and Note-taking Awareness Check

DIRECTIONS: Please put an "X" in the appropriate box.

	Yes	No	
1.	❐	❐	The keys to learning from lectures include active listening and good note-taking.
2.	❐	❐	When I hear material, I am listening.
3.	❐	❐	I am addicted to the fatal belief that I can listen to two things at once.
4.	❐	❐	My notes are easy to read and understand.
5.	❐	❐	I copy information the instructor writes on the board.
6.	❐	❐	I try to write statements summarizing what the instructor is saying.
7.	❐	❐	I revise and review my notes within 24 hours.
8.	❐	❐	If I do not understand my notes or they seem incomplete, I ask another student for help or see my instructor.
9.	❐	❐	I use abbreviations to accelerate my note-taking.
10.	❐	❐	I always believe I can gain new knowledge from a lecture.
11.	❐	❐	I am able to follow the instructor's line of reasoning in presenting ideas.
12.	❐	❐	I can narrow my focus when I listen to a lecture, so that other student activities, both in and out of the classroom, do not serve as distractions.
13.	❐	❐	I consider students' comments and questions to be an important part of the lecture, and I listen to them, preparing an answer to their questions.
14.	❐	❐	Even when I don't agree with the instructor, or I don't understand some part of the lecture, I remain an active listener.
15.	❐	❐	I make up test questions from my notes.
16.	❐	❐	An important function of notes is their availability for use for later review or study.
17.	❐	❐	Organization is important in note-taking.

Positive answers to the questions in the Awareness Check (except items 2 and 3) suggest that your note-taking skills resemble those of other successful students. Few people, however, are this efficient with their note-taking skills.

As you read this chapter, begin to incorporate some of the listening and note-taking techniques that are new to you in your classes and when you do your assignments.

Introduction

Learning to listen and take useful notes requires practice. As these skills are developed for use in the academic setting, notes become valuable in helping the student to understand the nature and purpose of academic lectures and other resources. Notes are used to help the student establish an active learning attitude. They direct the student's attention to getting meaning out of lecture material, accompany the studying process, and provide a record of material for further study and review. As the student listens, abstracts, organizes and condenses material, he/she is working to integrate and assimilate new information.

Listening

In the academic environment, you are continually exposed to lectures. If you wish to learn the information that is being taught, you must know how to listen. Listening is a neglected communication skill. It is an active process in which you transform the material that you hear into meaning for you. Allowing words to pour into your ear is not listening. You must organize material, relate it to your experiences, and make it part of yourself.

Active listening is effective listening

You can learn to listen more effectively.

Research indicates that the following statements are true (Carmen and Adams, 1972).

1. **We listen in spurts.** Your attention wanders so that you listen intently for 30 seconds or so, tune out for a short time, and then return. You are usually not aware this is happening.

2. **We hear what we expect to hear.** Your prejudices, past experiences, expectations, and beliefs determine what you hear. You tune out what you do not want to hear.

3. **We do not listen well when we are doing other things**.

4. **We listen better when we are actively involved in the process.** When we listen to satisfy a purpose, we hear more and better.

You may first need to identify the instructor's purpose in lecturing. Is it to show you how the course material relates to your own life, or to help you solve a particular problem? Is it to discuss and raise questions, and to demonstrate certain trends? Is the goal to encourage you to think critically and analytically? The instructor often states the purpose of the lecture on the syllabus as well as at the beginning of the lecture.

Identify purpose of listening

Active listening is a critical skill involving much more than just passively hearing the words someone else has spoken.

Restate

There are several active listening techniques. One is to restate or paraphrase in your own words what you have just understood the other person to say. In the classroom, this technique is used to check your understanding of what is being said by the instructor or other class members. In the employment situation, restating is a powerful skill to ensure accuracy and precision. Paraphrasing also helps to demonstrate your interest in what the instructor is saying, and encourages the speaker to explain more fully.

Paraphrase

Watch for non-verbal clues

The second active listening technique is to respond to non-verbal clues. Observe and understand the impact of words, watch for non-verbal clues when messages are delivered and received. Responding to non-verbal clues helps to get more information.

Summarize key points

A third active listening technique is to summarize the key points that have been made during the course of the lecture. Summarizing includes all key points that have been made and helps to keep you aware of what is important as well as to illustrate what has already been covered.

Ask questions

A fourth active listening technique is to ask questions that will help generate responses you can listen to. Use open-ended questions which will help when you need further information or explanation. Open questions begin with "what, how, who, when, and where."

Listen carefully for key points which are often given in the first few minutes and the last few minutes of class and meetings.

Note-Taking

Any effective study system must include skills for note-taking. An organized, planned approach to the lecture helps to increase on-the-spot learning, maintain a longer attention span during the lecture, boost retention of the material, and have valuable notes for later study as students review and at exam time.

Note-taking is a very real, immediate and practical college success skill. As a beginning college student, you will discover that the lecture method of instruction is frequently used by professors.

Taking notes helps the student remember what the instructor said during the lecture, and utilizes many of the sensory skills. You use your auditory skills as your ears take in information through careful listening; your visual skills as your hands and brain transform the words and ideas into a visual form that can be read, reread and reviewed. You might be aided by diagramming the information presented, highlighting, starring, and/or drawing circles around important/key points. Writing information down forces students to encode it, creating a deeper impression on their brains.

Look, listen, and write

Concentrate

Another important reason for note-taking during lectures is to help you concentrate in the classroom, to focus and to be an active listener.

Prepare

Good note-taking skills are essential in helping you to prepare for tests. Frequently, studying for a test consists of reviewing and memorizing information from lecture notes. It is important that these notes be complete, readable, accurate, and well-organized. In addition, notes are often a source of valuable clues to what information the instructor thinks is most important, and therefore what may be included on exams.

Readable, accurate, well-organized

Notes often consist of information that is not found elsewhere. Instructors frequently use lecture time to explain concepts introduced in the text, and to add professional examples the student can use in understanding this information.

Your ability to listen and take notes is closely-linked to how well you will do in class.

TAKING THE RIGHT KIND OF NOTES: DEVELOPING YOUR LISTENING AND NOTE-TAKING SKILLS

Many note-taking systems have been developed. The Cornell system (Pauk, 1984) can be applied to all lecture situations. Its goal is simple efficiency. Every step is designed to save time and effort. There is no rewriting in this system. Each step prepares the way for taking the next natural and logical step in the learning process.

Phase I: Before the lecture (adapted from Long, 1992)

1.   Take a few minutes to review your notes on the previous lecture, to provide continuity with the lecture you are about to hear.

2.   Divide paper in half, preparing to take notes on one side.

Phase II: During the Lecture

1.   Record your notes on one side of the paper, completely and clearly enough so they will still have meaning for you long after you have taken them. Strive to capture the main points of the lecture, the general ideas rather than illustrative details. Do not be concerned with developing an elaborate formal outline using Roman numerals, capital letters, numbers, etc. Subtopics under main points can be indicated with numbers or simply with a dash placed in front of each subtopic.

2.   Write so that you will be able to read the material. This may mean practicing a form of printing or developing a system of abbreviations. Be careful to use only those abbreviations you are familiar with.

Phase III: After the Lecture

1.   Consolidate your notes as soon as possible after the lecture by reading through them to clarify handwriting and meaning. Underline or box in the words containing the main ideas. Restructure the notes by reading them and then jotting down key words and key phrases from the lecture on the left side of the paper. This procedure helps you to recall the lecture. The process of writing summarizing words and phrases helps to fix the information in your mind.

Consolidate Restructure Summarize

2.   Cover your notebook page so that you are looking only at the key words and phrases. Use the jottings as cues to help you recall and recite aloud the facts and ideas of the lecture as fully as you can in your own words.

Uncover the notes, then verify the accuracy of your work. This is known as using your skills of recall. Recall is a skill you will need to demonstrate on your examinations.

Test your recall of information

3.   Review your notes. Work on recalling the contents of the lecture. Meeting with other students in the course is another profitable way to review your notes. Not only can you review the main points of the lecture with each other, you may also discover that other students have heard points that you haven't. Perhaps they saw the relationship of points to each other differently than you did. Input from other

students can be one more source of information for you as you give meaning to what you are studying.

Reduce ideas to concise summaries as cues for **Reciting, Reviewing, and Reflecting.**	Record the lecture as fully and meaningfully as possible.

In Summary:

1. **Record.** During the lecture, record in the main column as many meaningful facts and ideas as you can. Write so that you can read your handwriting.

2. **Reduce.** As soon after the lecture as possible, summarize (reduce) these ideas and facts concisely in the recall column. Summarizing clarifies meanings and relationships, reinforces continuity, and strengthens memory. It is a way of preparing for examinations gradually and well ahead of time.

3. **Recite.** Now cover the main column. Using only your jottings in the recall column as clues, state the facts and ideas of the lecture as fully as you can, not mechanically, but in your own words, and with as much appreciation of the meaning as you can. Then, uncovering the notes, verify what you have said. This procedure helps transfer the facts and ideas to your long-term memory.

4. **Review.** Spend time each week in a quick review of your notes. You will retain most of what you have learned, and you will be able to use your knowledge more effectively.

Exercise 6.2

Choose two pages of notes from one of your academic subjects. Read your notes and fill in the recall clues or formulate questions that would help you study and learn the notes for a test.

Recall Clues and/or Questions

_____

_____

_____

_____

_____

_____

_____

_____

_____

_____

_____

_____

_____

_____

_____

_____

_____

_____

_____

_____

_____

Professors expect students to actively listen and participate by asking questions.

Good note-taking is a powerful skill. You will be motivated to enhance this skill when you find your notes are useful in preparing for exams. Good notes may save you hours of study time.

When you take notes on a lecture, you must be well-prepared to focus on the information being presented, and to use a technique that will allow you to receive large amounts of academic material efficiently.

Poor note-taking wastes a lot of time and is essentially useless. A student must be able to read his or her notes, to develop study questions from his/her notes, to use key words from the notes to prepare essay and multiple-choice practice questions, and to feel that he/she can achieve mastery of the material presented in the classroom from a review of his/her notes.

Tips for Strengthening Your Note-Taking Skills

Tip 1:   Organize and plan your approach to note-taking. What will you write with?

Come to class prepared. Have a pen or pencil to write with. Black ink is usually best because it is a dark color which is easy to see and does not fade into the page. Ink lasts longer than pencil markings. A student is more receptive to studying when the ink is readable, not very light and fading.

What will you write on? Where will you keep your notes?

Have a separate spiral notebook for each subject area. Leave the front pages clean so that you may develop a table of contents for your notes as the semester develops. Number all other pages in the notebook so that you will have an orderly progression to your notes. If it is more comfortable, use a sectioned ring binder to which you can add loose-leaf pages for each subject. Again, keep a separate loose-leaf for each subject.

Leave space in your notes to add information and explanatory details.

Tip 2:   Take an active role in note-taking. Be an ACTIVE LISTENER.

Set up questions to keep yourself in the lead. Turn your reading and your instructor's lecture titles or opening sentences into questions. These are not questions that you ask your instructor, but ones around which you plan your listening. Make up your own questions, then listen for the answers.

Use questions

Pay attention to the instructor's organization, and his or her major points. Then jot down the basic ideas as you grasp them. Get in the habit of listening for major points and conclusions. Identify main ideas and the connections among them. Identify those general assertions that must be supported by specific comments. Identify the specific information (supporting details). Information presented to you in a series or a sequence is frequently worthy of note, (e.g.: "There are four reasons this occurs, . . ."). Active listeners pay attention to what they hear and try to make sense of it.

Jot down basic ideas

Ignore distractions that compete for your attention. Keep your focus on the material the lecturer is presenting.

Tip 3:   Develop your own abbv. for commonly used wds.

*	important
esp	especially
?	question
vs	against
. . .	and so on

Your own shorthand style will save time and make sense later when you are reviewing your notes. For example, psychology may be abbreviated as psy. You are already familiar with many abbreviations: days of the week, months of the year, states, college courses, etc. You also have many abbreviations you can use from your math studies, = (equal to); > (greater than); < (less than).

There are times when you shouldn't use shorthand. When you are given a precise definition, you will want to make sure you record exactly what is presented. This is also true when you are given a formula or an example of an application of the formula.

Tip 4:   Identify the main ideas.

Lecturers sometimes announce the purpose of a class lecture or offer an outline, thereby providing you with the skeleton of main ideas and details. Use this information to structure your notes, identifying the major points and the details that support them.

During the lecture, there are many clues which indicate that some of what is said is more important than other information. Some lecturers change their tone of voice, stamp the podium, or repeat themselves at each key idea. If your instructor emphasizes the same information repeatedly, the instructor feels this information is important. Chances are good that you will see these topics again on the exam.

Some lecturers ask questions to promote classroom discussion. This is a clue to what the lecturer believes is important. Identify the theme of the question.

Tip 5:   Bring your textbooks to class.

Many instructors refer in their lectures to information in the textbook. Sometimes they will ask you to do an exercise from the text, or review the interpretation of a graph or visual material accompanying data presentation. Some instructors lecture on material contained in the book and supplement this material. If you have your text in class, you will be able to follow along and note important material.

Tip 6:   Make a personal commitment to learn and use good note-taking skills.

Assume personal responsibility

A positive frame of mind will strengthen your motivation to be an active listener. Assume that you will learn something useful, that you will expand your knowledge, and that your understanding of the course will increase.

Tip 7:   Write down what the instructor puts on the board.

If a professor takes the time to write points on the board, you need to give that information special consideration.

Tip 8:   Write down and use the dictionary to learn unfamiliar words.

Ask the instructors in class what unfamiliar words mean. Use these new words in your vocabulary.

Tip 9:   Be an ACTIVE PARTICIPANT.

Summarize statements of information presented in the lecture in your notes or to the class-at-large. **Take notes with a purpose.**

Summarize material

Don't be afraid to ask questions. If you don't understand some of the material, it's very likely that your classmates have similar concerns.

When you realize that you have missed an important point, ask the instructor to repeat it. If you don't understand what is being said and need time to dwell on it, leave a space in your notes and put a question mark (?) in this place. Fill this void in your notes by asking a classmate or the professor prior to the next class session.

Ask questions

You may find graphs, charts, and drawings to be helpful. When these are used to illustrate a point, make your own sketch of what the instructor has presented.

Tip 10: Do not rely on a tape recorder for note-taking.

If you use a tape recorder, do not allow this to encourage you to become a passive learner. You still need to write the main points the instructor is lecturing on.

Tip 11: Recall

Create a recall column in your notes. This remains blank while you take notes during class. It is, however, used within 24 hours of the note-taking to review and synthesize your learning of the material. Write the main ideas and key information covered in the notes in the recall column.

Tip 12: Recite

Use key words or phrases highlighted in the recall column to recall and recite out loud what you understand from the class notes. This summarization of your notes can then be used to prepare for test-taking.

TIP 13: Review

Before lectures begin, review notes from the previous day. This is a "warm-up" to help your mind focus on the material to be covered and to prepare you to think critically during the lecture.

Re-writing helps recall

Notes are to be used frequently. Within 24 hours after the lecture, go through your notes and complete any information which might have been recorded hastily or with the intention that you would provide more detail after class. Make sure concepts are clear and understandable. If not, read your book; check your notes against the text, especially if you missed some main points while writing them down; speak with other students; or check with your instructor. You may also bring these questions up with the professor at the beginning of the next class.

Compare notes with other students and discuss for better retention and understanding.

Work on condensing thoughts, ideas or facts into a few words or phrases which will be meaningful to you at a later date.

Tip 14: Find out who's the best note-taker in your class.

Compare notes, borrow notes, restructure notes. Should you miss a class, make copies of these notes.

Tip 15: Build test questions from your notes.

Once you have identified the key points in a lecture, you can identify exam answers by making up your own set of exam questions. This is exactly what the instructor does in making up an exam, giving most of the same questions—and the answers, too. In a study conducted at one eastern college, a group of students was asked to use this study method. It was found that up to 80 of the actual exam questions were among the key point questions the students had made up ahead of time. The grades of these students were 10 points higher than students not using this method of study (Olney, 1991).

After you have developed all of your questions, use your notes to highlight the answers for each question. If you find an answer is incomplete in your notes, fill in the necessary detail. You may want to transfer questions and their answers to note cards for easy and "portable" review.

Tip 16: Apply your note-taking skills as you study and mark your books.

Use your pencil as you read.

Underline important points.
Write notes in the margins.
Draw arrows connecting important material.
Circle material you want to focus on.
Find a system of marking your textbook that is right for you.

These activities help you to be actively involved in note-taking. Activity such as this forces you to focus on the material and to concentrate.

Active reading and note-taking helps you to search for what is important. Your markings serve a similar purpose to the recall column while the underlined text provides supporting detail.

Exercise 6.3

COMPARING NOTES

Pair up with another student and compare class notes from your core courses such as humanities, social environment, psychology, natural sciences, and this class. Are your notes clear? Can the other student identify important points from your notes? Do you agree on what is important? Take a few minutes to give feedback and explain to each other your note-taking system.

_____

_____

_____

_____

_____

_____

_____

Exercise 6.4

Note-taking is a critical skill you continue to develop throughout your academic career. It is essential to survival in college. Compare your present style of note-taking with the approach suggested in this chapter. Where are you having the most success in taking satisfactory notes which are a good resource when you are studying for exams. Why? Discuss at least two of the note-taking suggestions you will implement immediately to improve your note-taking skills.

_____

_____

_____

_____

_____

_____

The benefits of good note-taking are many. These include more interest in the material, an enhanced ability to apply the material in many situations, better retention, and improved notes for later study.

It does take time and self-discipline to use a note-taking study system. As you progress in college, you will be forming new habits with regard to your listening skills and your attitude when attending a lecture.

Journal Questions/Activities

1.   What did you expect to learn from this chapter?

2.   (a) What problems do you have when you take notes?

     (b) How can you eliminate those problems?

3.   How can you be a better student in class?

4.   Identify at last two things you learned about yourself from this chapter.

5.   Did this chapter teach you what you expected? If not, what would you have liked to learn?

Summary

Listening and note-taking are critical skills for academic success. Notes provide a useful and convenient record for study and review. Properly used, they are an important reference component and study supplement. With clear and complete notes, a student may identify key lecture points and integrate academic information as well as develop and practice exam questions which focus on the mastery of academic material. Use your notes to learn more, to gather information, and to ask meaningful questions. Remember, learning is a participatory sport.

Summary Exercise 6.5

1. Why is it a good practice to make a habit of reviewing all of your notes within a 24-hour period?

    _____

    _____

    _____

2. What is the purpose of a recall column?

    _____

    _____

    _____

3. List several verbal and visual cues the instructor may use to indicate certain information is important and belongs in your notes.

    _____

    _____

    _____

4. How do you want to improve your note-taking skills?

    _____

    _____

    _____

5. What factors would you consider important to take notes on when researching careers?

    _____

    _____

    _____

    _____

References

Carman, R. & Adams, Jr., W. (1972). *Study Skills: A Student's Guide for Survival*. New York: John Wiley and Sons.

Lawson, H. H. (1989). *College Bound Blacks*. Dubuque, Iowa: Kendall/Hunt Publishing Company.

Long, Kenneth F. (1992-a). "Listening and Learning in the Classroom," Chapter 8 in Cooper, C. (Ed.), *Keys to Excellence* (first edition), Dubuque, Iowa: Kendall/Hunt Publishing Company.

Long, Kenneth F. (1991-b). "Listening and Learning in the Classroom," Chapter 4 in Gardner, J. & Jewler, A. J. *Your College Experience: Strategies for Success*. Belmont, California: Wadsworth Publishing Company.

Maring, G., Burns, J., & Lee, N. (1991). *Mastering Study Skills*. Dubuque, Iowa: Kendall/Hunt Publishing Company.

Miami-Dade North Campus SLS Orientation Files (1992).

Olney, C. W. (1991). *Where There's a Will, There's an* A. Chesterbrook Educational Publishers, Inc.: Paoli, PA.

Pauk, W. (1984). *How to Study in College*. Boston: Houghton Mifflin.

Chapter Seven

Memory

Max Lombard, Ed. D.
Carol Cooper, Ed. D.

By permission of Johnny Hart and Creators Syndicate, Inc.

Exercise 7.1

Memory Awareness Check

DIRECTIONS: Please place an "X" in the appropriate box.

	Yes	No	
1.	☐	☐	You forget because you did not understand the information.
2.	☐	☐	You forget because you did not associate the information.
3.	☐	☐	Without memory you would be condemned to respond to every situation as if you had never experienced it.
4.	☐	☐	Short-term memory holds new information for approximately nine seconds.
5.	☐	☐	It is not necessary to review immediately after receiving new information.
6.	☐	☐	Acronyms and chunking are mnemonic devices.
7.	☐	☐	You receive information only through your senses.
8.	☐	☐	Learning is really a passive not active process.
9.	☐	☐	The process of memory consists of three stages.
10.	☐	☐	One of the five R's to move memory from short-term memory to long-term memory is receiving.
11.	☐	☐	Short-term memory is active memory.
12.	☐	☐	Spaced study is efficient use of study time.
13.	☐	☐	You should learn from the specific to the general.
14.	☐	☐	It is more efficient to use one sense at a time when reviewing.
15.	☐	☐	New information needs to be associated with old information.

Introduction

As a freshman, you will be responsible for learning from your teachers, textbooks, lectures, periodicals, libraries, media presentations and extra-curricular activities.

Studying and learning are hard work. Your success in College will depend on your determination and your willingness to apply yourself.

This chapter will provide you with terms and concepts you will need to know in order to understand why you forget or remember information. You will learn how memory works and become aware of a number of strategies to move memory from short-term memory (STM) to long-term memory (LTM). Remember, no one can learn for you. If you have the motivation and desire to learn, you will successfully complete this class and, eventually, your program of studies.

You need to be mentally prepared to accept new information

Why You Forget

Your ability to forget information is virtually unlimited, given limited practice.

A.   You forget because you may not have RECEIVED the information.

B.   You forget because you have a weak IMPRESSION of the information.

C.   You forget because you did not UNDERSTAND the information.

D.   You forget because you did not pay enough ATTENTION to the information.

E.   You forget because you never STORED the information in the first place.

F.   You forget because old information INTERFERES with new information.

G.   You forget because you did not REHEARSE, REVIEW, RECITE, REPEAT and REPEAT AGAIN the information.

Review frequently to prevent forgetting

H.   You forget because you did not ASSOCIATE or LINK or CONNECT the new information to familiar or old information.

I.   You forget because you were not interested in the information.

J.   You forget because the new information CONTRADICTS your prior beliefs.

K.   You forget because you may not have used RELEVANT strategies to help you retain the information.

L.   You forget because you have not ORGANIZED the information.

M.   You forget because you may REPRESS the information.

How Memory Works

> Without memory you would be condemned to respond to every situation as if you had never experienced it.
>
> You would be incapable of thinking and reasoning.
>
> Think of it.
>
> You would have to start from scratch every time.

The process of remembering consists of three stages.

3 R's   A. RECEIVING   B. RETAINING   C. RETRIEVING

RECEIVING or acquiring or inputting

You first have to receive the information. You cannot remember what you have not received. The information first must get into your memory.

RETAINING or maintaining or storing

You have to retain the information received. Memory must last inside the brain for long periods.

RETRIEVING or remembering or recalling

You have to be able to recall the information retained.
You must be able to reach into the memory storage and find the precise piece of information you need at the moment.

Thus, the three R's of remembering are RECEIVING, RETAINING and RETRIEVING. Now look at each stage and determine how it functions.

The First Stage of Remembering

RECEIVING or acquiring or inputting.
You receive information through your five senses.

The Five Senses

Hearing Auditory	You hear information	Concert Lecture Radio Discussion
Visual Sight	You see information	Television Blackboard Movies Class Notes Books Periodicals
Motor Tactile Touch	You feel information	Feel an object Take notes Play an instrument Ride a bike Skate Walk Run
Smell Olfactory	You smell information	Flowers Meals Perfumes Odors
Taste	You taste information	Wine Food Medicine

Just as right- and left-handed individuals give preference to one hand, most people give preference to a sensory reception. Since you receive information and learn through your senses, you must determine which one of your senses you favor when receiving information.

If you favor hearing, you probably will do well in lectures or discussions. On the other hand, if you favor sight, you may receive better in visual presentations.

The fact is that your brain receives and processes information through your five senses. Without your senses, your brain could not receive information.

Think of it. If you could not see, hear, feel, smell or taste, you would not perceive (be in touch with) the environment surrounding you. You could not react, understand or learn from it.

On the other hand, the more senses you use on new or old information, the better the reception and eventual understanding and learning.

Research shows that most people will retain about 15 percent of the information they hear (the auditory sense), but they will retain about 85 percent of what is presented to them visually through the following: diagrams, writings on a blackboard, films, pictures, even words in a printed page. Research also shows that involving motor senses through touch significantly strengthens and deepens the learning accomplished by the other two senses. Thus using the sense of touch to write information down in addition to hearing and seeing the information creates a deeper impression in our brains.

It makes sense to use your senses

For example, while listening to a lecture, look either at the speaker or, if the speaker is using the blackboard, newsprint, an overhead projector, etc., look at what is presented. Taking notes during the lecture includes the sense of touch and sight. For a visual learner, the strategy of note-taking during a straight lecture helps one to see what the lecturer is saying. The person also uses the sense of touch when writing.

Regardless of your dominant sense, the more senses you use, the better the reception. TO BECOME A BETTER RECEIVER, YOU MUST ENGAGE AS MANY OF YOUR SENSES AS POSSIBLE. You must use a combination of senses—for instance, writing and reciting or hearing with seeing, exercising, and repeating.

Understanding

When you RECEIVE information, you need to understand it. Information becomes much more difficult to retain, and eventually to recall, if you have not understood it in the first place.

In order to understand new information—information that is unfamiliar—you must ask questions, relate the new information to old or familiar information, and discuss the new information with a fellow student.

Association

The extent to which we remember a new experience has more to do with how it relates to existing memories than with how many times or how recently we have experienced it.

Morton Hunt

You can remember any new piece of information if it is associated to some thing you already know or remember

When you RECEIVE new information you need to personalize it, to link it to learned or familiar information.

Having received the new word "dilapidate," which means to bring into a condition of decay or partial ruin by neglect or misuse, you may associate the new word to an old house with which you are familiar. That house is really dilapidated. It has been neglected. You may associate the new word with the Roman Colosseum which is in ruins. The Colosseum is dilapidated. You may remember that "lapis" means stone and that dilapidate means to take stones down, to tear down. The Colosseum has been torn down. It is dilapidated.

When you relate or associate new information to your experience, you create connections.

Learning Is an Active Not Passive Process

Be an active participant during a learning experience.

During lectures, ask questions if you do not understand a point. Ask for clarification. See the instructor after class or during his/her office hours.

Discuss lecture notes with your fellow students and compare notes.

If you become bored during a presentation, become active, ask questions, guess what the presenter is going to say next, and sit up straight. Boredom is impossible if you are active. Boredom is the product of inactivity. When you are inactive, you are not RECEIVING.

Strategies to make learning active

An Active Learner Receives the Information

Be in the Here and Now

During the lecture or presentation, pay attention and be observant. Be focused in the moment.

Concentrate on the moment by removing all distractions from your mind. What distractions are you experiencing right now?

When you are not attending, actively listening and observing, when you are not focused on the moment, you usually are removed. You are either in the past or the future, not in the present. Remember that the past is gone and the future has not yet come. It is only in the moment, in the present, that information is being received. For example, when you are listening to a lecturer, reading a book, reading your notes or preparing for a test, there is always the tendency to drift away, to daydream, to wander away from what is happening right now, this moment, and either go into the past or the future. You may wander to an upcoming date, a ball game, finances, family issues, next class period, etc.

By doing this, you are not attending to what is going on at the moment, thus not receiving the information.

The best strategy you can use when wandering away from the present is to become aware of what you are doing and then gently talk yourself back to the present and the activity at hand. For instance, while you are listening to a lecture on modern art, you drift away and start thinking about your date and what you plan to do that evening. When you become aware of what you are doing, you should tell yourself, "Yes, that is nice, but I am looking ahead and I need to return to the present and RECEIVE the information at hand. If I choose not to do this, I will not receive the information I need to pass this course."

Become aware of the present

REVIEW

1.   You receive information through your five senses. The more senses you use, the better the reception.

2.   You need to understand the information you receive. Ask questions, discuss it with classmates, and see the instructor.

3.   Be an active learner. Be present, attentive, observant and aware. You need to become involved.

4.   Stay in the present, the here and now, rather than wander into the future or the past. By staying focused in the present, you will receive information.

5.   You need to associate new information, personalize new information, and link it to learned or familiar information.

The Second Stage of Remembering

RETAINING or maintaining or storing

Once you have received information, you need to retain it. In order to do so, you must take some specific action to prevent forgetting the newly-received information.

Information Not Used Is Soon Forgotten

USE IT OR LOSE IT!

**The Curve of
Forgetting**

(The Ebbinghaus Curve)

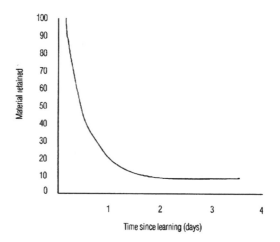

Figure 7.1

Without reviewing new information, you will forget 80% the first day. To retain information you received and recall it later, you must move it from your short-term memory (STM) to your long-term memory (LTM). Before trying to remember this information, make sure that you have properly organized the material the way it was intended.

Short-term memory is active memory. It holds the contents of your attention. Short-term memory is very short-lived.

Short-term memory

Attention determines what goes into short-term memory. People are selective about the focus of attention; thus much of what they are exposed to will never even enter short term memory. Short-term memory can hold only six to nine items at a time.

For example, when you look up a new phone number or hear a new name or receive directions, this information will decay within 15 to 30 seconds unless you consciously process the information into short-term memory. Unless you use the strategies offered in this chapter, you will forget them.

Exercise 7.2

Answer the following questions without looking up the information you have just read.

1. How do you receive information?

   _____

2. Why is it important to stay in the present (the here and now) during a lecture or when you do your homework?

   _____

3. Draw a graph that represents the curve of forgetting.

4. List the reasons you forget.

   _____

5.   Define short-term memory

_____

6.   What is meant by "associate new information."

_____

Reviewing or using a piece of information acts as a rehearsal, thus strengthening the specific memory trace.

REVIEW this information and try this exercise again tomorrow. RECITING a new phone number or name activates another sense and reinforces the short-term memory. If you REPEAT the new information, write it down and use it a number of times, you may move it from your short-term memory to your long-term memory. If you want to remember new information for more than a short time, you must move it to your long-term memory which holds information for a limitless time. STM is the gateway through which information may pass into LTM.

Long-Term memory

Long-term memory holds information for long periods. In many cases, memory in LTM is stored for a lifetime.

For example, you probably can recall your home phone number and address, your ability to drive a car, the first president of the United States, the current president of the United States, the multiplication table, the name of this community college, the name of the president on a five-dollar bill, the meaning of the word procrastination, and the furniture in your bedroom.

If you think about it, you have stored a great variety of information in your long-term memory.

The point is that the transfer of information from short term to long-term memory is crucial if you want to be able to recall the information at will.

In order for learning to take place, and for you to be able to recall information at a later date, you must transfer information from short-term to long-term memory. You must take some specific action to retain information. The following are strategies you can use in order to solidify the information in your long-term memory.

The five R's to move memory from short-term to long term memory are

REVIEW, REHEARSE, RECITE, REPEAT, and REPEAT AGAIN.

You can go over recently-received information by rereading your notes or the chapter, discussing the new information with classmates, writing an outline, using 3″x5″ cards with questions on one side and answers on the other, and associating the new information with the old information.

Review

Remember, you can **review** the information by repeating it aloud, writing it, discussing it, and visualizing it. The more senses you use in the review process, the more effective the review. You should review immediately following the reception of new information. Review weekly, review before a test, and review again before the final.

By **reciting** the information aloud, you anchor the concepts by using two senses: **the physical sensation of speak**ing and by hearing what you are saying. You are also repeating the information. Recite the material until you can recall it without the help of your notes.

Recite

Just as actors have to **rehearse** the play in order to succeed on opening nights, you need to rehearse the information in order to succeed on the test. Rehearsing lets you know whether you have mastered the information. You can rehearse in front of a mirror, with friends, or with 3″x5″ cards. You can use pictures and various memory strategies.

Rehearse

By reviewing, reciting, and rehearsing, you are **repeating** the information. In courses such as accounting, chemistry, mathematics, biology, and anatomy, you may need to memorize specific terms. In other courses such as philosophy, ethics, history, and anthropology, you may have to memorize specific terms and dates, but you may also have to be able to paraphrase information. State it accurately in your own words, not the author's words. Only through this repetition will you remember the information. Think back to when you learned a new song. You tried to sing it over and over again until you knew the entire song. What about those commercials you find yourself singing? The advertisers played them over and over again until they figured you were familiar with their product. Then the ads tapered off.

Repeat and repeat again

You should use the 5 R's in order to move STM to LTM and eventually pass the final, pass this course, and finally complete your college program.

Practice strengthens learning

Other Strategies to Reduce Forgetting

Mnemonics are techniques, tricks, games, and rhymes that help you recall separate bits of information that cannot be associated, connected or linked in any meaningful way. You will likely remember your mnemonics if the are: humorous, peculiar, colorful, clearly relate to what you are trying to remember and are symbolic. Thus, if you have to learn the categories with which all living creatures are divided, the following mnemonic technique known as an acrostic may help you.

The mind insists on finding meanings and patterns

Acrostic is a mnemonic device by which you create sentences or phrases by using words that begin with the first letter of a series of words.

The first letter of the statement above the line may help you remember the terms below the line.

King	Phillip	Came	Over	From	Greece	Singing	Vigorously
Kingdom	Phylum	Class	Order	Family	Genus	Species	Variety

If you have to learn the twelve cranial nerves, try this technique.

The first letter of the statement above the line may help you remember the terms below the line.

On	Old	Olympus	Towering	Tops	A
Olfactory	Optic	Oculomotor	Trochlear	Trigeminal	Abducens

Finn	And	German	Viewed	Some	Hops
Facial	Acoustic	Glossopharyngeal	Vagus	Spinal	Hypoglossal

The following is another mnemonic gimmick.

There are two varieties of camels, the one- and the two-humped variety. One is called a Dromedary and the other a Bactrian. In order to determine and remember which has one or two humps, you only need to place the first capital letter of each camel on its back.

$$D = \cap \quad \text{and} \quad B = \text{ɱ}$$

Now you can easily determine that

$\cap$ = Dromedary has one hump
B = Bactrian has two humps

Exercise 7.3

Invent an acrostic that will help you recall the planets in the solar system in order of their distance from the sun.

A former student, Mirna Lucho, came up with this acrostic:

My	Vote	Entitles	Magic	Johnson	Some	Upper	Nike	Pumps
Mercury	Venus	Earth	Mars	Jupiter	Saturn	Uranus	Neptune	Pluto

Now create your own.

| Mercury | Venus | Earth | Mars | Jupiter | Saturn | Uranus | Neptune | Pluto |

Visualize by creating a picture or image of a concept, idea or passage that you want to remember. You could do this by forming pictures in your mind or by drawing pictures, graphs, diagrams or cartoons on paper.

Right-handed individuals process most visual information in the right hemisphere of the brain. Verbal information is generally processed in the left hemisphere. When you visualize, or draw a picture, graph, diagram, or outline that relates to a concept, idea, fact or notion, you are using both hemispheres or two areas of your brain, thus increasing your chances to retain and consequently to retrieve the information. The more active the picture, the clearer the picture. The more unusual or more exaggerated the picture, the better it will help you recall the information.

Visualization can be used effectively to link terms, names, and concepts with places, areas, and times. In anatomy, you may visualize the brain or the skeletal system. In geography, you may visualize locations. For example, visualizing this chapter, you may picture a zoo. The locator map at the entrance represents short-term memory; the rest of the zoo represents long-term memory. Different areas of the zoo can represent stages of the process of memory. For instance, the ticket-taker who receives you (Receiving), the fences that hold the animals (Retaining), and the loud calls of the animals (Retrieving), can help you remember the process. Now create your own visualization.

Pegging is a memory strategy in which you visualize a number of locations or objects in a particular place, etc.

Chunking is the process of organizing information into sets. Since short-term memory can attend to and remember only about six or nine pieces, item sets, or chunks of information at a time, a good learning strategy is to combine sets of information. Thus, it is easier to remember a social security number if it is chunked 374-81-9350 rather than to remember 3-7-4-8-1-9-3-5-0. The reason for this is that you have reduced nine sets to three sets.

Acronyms are words created from the first letter or the first few letters of the item or words on a list. For example, HOMES reminds you of the five great lakes—Huron, Ontario, Michigan, Erie and Superior. Another example of an acronym is SCUBA—Self-Contained Underwater Breathing Apparatus.

Exercise 7.4

The following are acronyms: SOAP, CLAST, RADAR, and NASA. Do you know what the letters stand for?

_____

_____

_____

_____

Rhymes and Tunes is a technique used to teach children basic facts and to condition adults into buying products.

For example, you probably will remember the following rhymes: "Thirty days hath September, April, June, and November;" "One, two. Button your shoe. Three, four. Shut the door..." and "In fourteen hundred and ninety-two, Columbus sailed the ocean blue." You may remember "Pepsi Cola hits the spot. Twelve full ounces is a lot" or "Uh-huh. You got the right one, baby," and even "A pint is a pound the world around."

Senses   **Reviewing with a Physical Activity** will force you to use more senses. Remember, the more senses you use, the more efficient your reception. Review while you perform a physical exercise such as jogging, walking, swimming, or bicycling.

Focusing is tuning in to the subject. Only when a camera is focused does it show a clear picture with all its details. When your mind is focused on the subject at hand and not distracted by daydreaming, watching television, or listening to a conversation, you receive a clear picture of the subject.

Clear Picture

Paraphrase is the ability to substitute your words for someone else's without altering the original meaning of the statement or concept. Since understanding is essential for in depth remembering, your ability to paraphrase exposes your depth of understanding.

Exercise 7.5

Paraphrase the following statement:

> *They that can give up essential liberty to obtain a little temporary safety deserve neither liberty nor safety.*
>
> *Benjamin Franklin*

Organizing information in a meaningful pattern allows you to develop the necessary links among the information received.

Meaningful Pattern

Learning from the General to the Specific allows details to become more meaningful.

For example, you can get a broad picture of the entire text by scanning or previewing the table of contents, headings, pictures, exercises, and introductions to each chapter. This practice will give you a good overview of the entire text. It will give you the big picture which later will help you study each chapter in detail. You should form a broad overview of the subject before beginning to learn the details. If you spend some time in learning the big picture, the details will be easier to remember.

Spaced Study is efficient use of study time. Long study sessions are not as effective as shorter sessions. You will not accomplish as much by studying for five to six hours at a time as if you limit your sessions to one or two hours with rest or change-of-pace intervals. You can accomplish more by taking regular breaks. **You can learn more at a faster rate if you work for a half hour or an hour, then take a break.** When you spend long periods studying, you tend to wander off or daydream.

Distribute Learning

Stone
What does the root LAPIS mean? Examples are Lapidarian, Lapidary, Dilapidate.

3″x5″ Cards are an effective learning technique in which you write questions on one side of the card and answers on the other side. When you need to review a chapter for a test or notes for a final, you can use the cards by yourself, with a partner, or in a group. The cards are similar to "Trivial Pursuit." You answer each question and check your accuracy by turning the card over. Do this until you can answer all the questions without making a mistake.

Cramming

Cramming is a strategy you may use as a last resort—it is better than nothing.

Overlearn by going over it again just to make sure you absolutely know it.

Combine all the memory techniques that work for you. Just as using more than one sense improves your ability to remember, it is equally effective to use various memory techniques.

The Third Stage of Remembering

RETRIEVING, remembering, recalling or recognizing.

Consider two types of **retrieving**.

Recall and **Recognition**

Recall

When asked to RECALL information you are required to retrieve all the information from your memory.

Example: List and define four mnemonic techniques.

1.

2.

3.

4.

Recognition

On the other hand, if you are asked to RECOGNIZE information you are merely required to distinguish information presented to you.

Example: Match the mnemonic technique with its definition.

Mnemonic quiz

1. Chunking   A. A device by which you create sentences or phrases using words that begin with the first letter of a series of words.

2. Acrostic   B. A device in which you visualize a number of locations or objects familiar to you in order to remember a list of ideas or items.

3. Pegging   C. Words created from the first letter or the first few letters of the items or words on a list.

4. Acronym   D. A memory strategy in which you organize information into smaller sets.

Recall question

In objective questions that require you to fill in information RECALL is necessary; there are no hints, no clues, you must RECALL all the information.

Matching question

On the other hand, in objective questions that list a number of items in which you have to correctly choose or match items from a given list, you merely have to RECOGNIZE information which is presented to you.

Essay tests use questions such as: In relationship to you or other college students, what is the significance of David McClelland's study?

Given this question you must RECALL David McClellan's study as well as relate this study to you and other students.

More effort and preparation is necessary in order to prepare for a test or quiz which requires you to RECALL information.

By following the suggestions offered in this chapter, you will have successfully RECEIVED information using all five senses; you will have been an active learner; you will have remained in the present when receiving information; you will have associated new information to old and familiar information.

You also will have RETAINED the information by moving new information from STM (Short-term memory) to you LTM (Long-term memory) by reviewing, rehearsing, reciting, repeating and repeating again; by using mnemonic devices such as visualizing, chunking, acronyms, pegging, acrostics; by combining review with physical activity; and by paraphrasing, organizing, spacing study and using 3″ x5″ flashcards.

Having successfully RECEIVED and RETAINED information, you now will be able to RETRIEVE information at will. You will pass the final exam in this course, and by applying the strategies offered in this chapter to all your courses, you will be able to succeed in your total program of study.

Tips on How to Improve Memory

The following tips will help you improve your memory, but you must be willing to make the necessary effort in applying them to your learning tasks.

Tip 1:   See the significance

Try to see the significance of what you are learning. Try to be interested in it, or at least in the value of remembering it.

Tip 2:   Give your attention

Give the material all of your attention. Be sure you have it right.

Tip 3:   Understand the material

Be sure you fully understand the material. Can you explain it to someone else so they will understand it?

Tip 4:   Make up your mind

Intend to remember the material. You have to make a conscious effort to remember the information.

Tip 5:   Have confidence in yourself

Be confident that once you put your mind to it, you will remember the information.

Tip 6:   Learn to associate

Associate new material with related facts you already know. When new material seems to disagree with previous learning, you will have a valuable handle for recalling it later if you can connect it to old information.

Tip 7:   Organize the Information

Organize the material so you can file it in its proper place in your memory. If you have organized carefully, remembering part of something will enable you to remember the rest. (When you reach through a door and grab the dog, it does not matter what part of the dog you catch. Be it a tail, ear, or leg, when you pull it through, the rest of the dog will follow.)

Tip 8:   Get the big picture

See how your learning is part of a larger whole.

Tip 9: Learn in segments

If there is a basis for doing so, divide and group your material into smaller segments or bunches. Information is best remembered when taken in in "little bunches."

Tip 10: Use it or lose it

Reinforce what you have learned through repetition and usage. You will not remember something you do not use.

Tip 11: Remember to recite

Test yourself repeatedly. RECITE from memory.

Tip 12: Distribute your learning

Study the material over a period of time in several sessions. Use spaced study or shorter study sessions rather than long sessions.

Tip 13: Try tricks of the trade to remember

Use mnemonic devices and try to visualize in pictures.

Tip 14: Make use of your senses

Use all your senses when reviewing: listen, read, recite out aloud, write it down, discuss it with others, practice and touch.

NOW DO IT!

 Journal Questions/Activities

1. Think back to the beginning of the term when you started this class. Write down at least one concept from each chapter that you believe you have learned without looking them up.

2. Using information from questions #1, share reasons why you think you remembered these concepts. Were there some important concepts you should have remembered but forgot?

3.   Do you forget personal events in your life the same way? Why or why not?

4.   What strategies from the information presented in this chapter will you use to enhance your ability to retain information?

Summary

This chapter covered the reasons you forget. You forget because you did not receive, understand, store, associate, use, and repeat the information. You also forget because you may have a weak impression of the information or the new information interferes with old information.

The process of memory consists of receiving, retaining and retrieving. You receive information through your five senses. You store information for a brief time in short-term memory and you must take some specific action in order to move information to long-term memory. You need to understand the information received and associate it, to be an active learner, and to stay in the present.

The more senses you use, the better the reception.

The five R's to move memory from short-term memory to long-term memory are review, rehearse, recite, repeat and repeat again. Other strategies to improve memory are mnemonics, acronyms, chunking, visualization, rhymes, tunes, reviewing with a physical activity, focusing, paraphrasing, organizing, spaced study, and 3″x5″ cards. Combining memory strategies that work for you is as effective as using more than one sense when receiving. Use the memory technique that works for you and do not be afraid to experiment with new techniques.

Your brain never loses anything! The trick is to learn and use strategies that will help you recall information.

Name _____   Date _____

Summary Exercise 7.6

1. Define "acronym" and give an example.

    _____

2. The process of memory consists of

    _____

    _____

3. List at least five reasons that cause you to forget.

    _____

    _____

    _____

4. List and describe three mnemonic techniques.

    _____

5. What is the advantage of being able to paraphrase?

    _____

6. Why is it effective to learn from the general to the specific?

    _____

7. Describe three memory strategies you use.

    _____

8. Long-term memory can take place if you

_____

9. How do you receive information?

_____

10. Define short-term memory.

_____

11. In this exercise you needed to RETRIEVE information. Were you forced to RECALL or just to RECOGNIZE the information? What is the difference?

_____

References

Baddely, Alan D. *The Psychology of Memory*. Basic Books, Inc.: New York, N.Y., 1976

Ebbinghaus, H. *Memory Trans*. Dover Publications. New York. 1913.

Ellis, Dave. *Becoming a Master Student*, 10th edition. New York: Houghton Mifflin Co., 1998.

Guthrie, E. R. *The Psychology of Learning*. Harper and Row: N.Y., 1952.

Higbee, Kenneth L. *Your Memory—How It Works and How to Improve It*. Prentice Hall, Inc.: Englewood Cliffs, N.J., 1977.

Hopper, Carolyn. Memory Principles. 2003. `http://www.mtsu.edu/~studskl/mem.html`.

Loftus, Elizabeth. *Memory*. Addison Wesley Pub. Co.: Reading, Mass., 1980.

Lucas, Jerry and Lorayne, Harry. *The Memory Book*. Ballantine Books, Inc.: New York, N.Y., 1975.

Luria, A. R. *The Mind of a Mnemonist*. Basic Books, Inc.: New York, London, 1968.

Norman, A. Donald. *Learning and Memory*. W. Freeman and Company: San Francisco, 1995.

Chapter Eight

The Art of Test Taking

Carol Cooper, Ed. D.

Name _____   Date _____

Exercise 8.1

Test Taking Awareness Check

DIRECTIONS: Please place an "X" in the appropriate box.

	Yes	No	
1.	☐	☐	Tests are given to determine your intelligence.
2.	☐	☐	Cramming is more likely to keep the test information fresh in our minds for the test.
3.	☐	☐	Students who pass tests are the most intelligent.
4.	☐	☐	Taking daily notes is a part of test preparation.
5.	☐	☐	There are basically only two kinds of formats for testing.
6.	☐	☐	Multiple choice tests usually consist of the stem and the distracters.
7.	☐	☐	Absolute words and qualifiers in test questions usually will tell you the answer.
8.	☐	☐	In an essay question when you are asked to do a contrast, you must explain the similarities and differences.
9.	☐	☐	Test preparation begins on the first day of class.
10.	☐	☐	One of the key reasons for test anxiety is that some students cram for exams.

Introduction

Testing is the age old method of finding out what students know about assigned course information and/or how well they have mastered a skill. Examinations force students into learning material and provide the instructor with feedback on how well they have been taught a subject and whether they need to modify their delivery of course information. Test results also provide students with information on how well they are progressing in the course. This information should tell students if they need to modify their method of studying in order to successfully pass the course.

Reason for tests

> Since testing is such an important college survival skill, you need to know just how it is done and what the rules are for succeeding.

Therefore this chapter will provide tips and strategies on how to effectively prepare for and take tests. It will also emphasize the importance of preparing for tests.

Taking tests is a four part process that requires a lot of work on the part of the student. Testing is a skill. Therefore, it must be practiced in order to learn it. You have to work at being a successful test taker just as you must practice and rehearse any other skill you want to learn. The four part process consists of the following:

Chapter objectives

- General Preparation
- Test Specific Preparation
- Taking the Test
- Reviewing After the Test

The four steps to becoming a successful test taker

You must be a smart student if you want to pass the test. The smart student is sufficiently motivated and understands why he or she must pass the test. Remember, it is not always the most intelligent, organized, or the best students who do well on a test and earn good grades. It is those students who have developed good study skills in test preparation and are committed to passing a course. These are generally the students who perform well at exam time. They are committed, time conscious and once again, organized and are willing to put in a sufficient amount of time in test preparation to make the grade. Here are some **tips** to help you understand this process.

Students who test well

Recommended website for test preparation and taking
`http://www.iss.stthomas.edu/tstprp1.htm`

General Preparation—Part I

Tip 1:   Know the planning rules for effective studying.

Questions to Ask Your Instructor

Start studying for exams from the first day of class. When talking with the instructor after you have received the course syllabus, be sure to ask the following questions:

a.   What are the major goals of the course?

b.   What kinds of exams do you normally give? Do you have any copies of some of your old exams I may use to help me prepare for your exams?

c.   How many units or chapters are usually covered on an exam if it is not already in the syllabus.

d.   Does the instructor expect you to remember general or specific detail, deduction and/or opinions?

e.   Will the tests normally cover class texts, lecture and other assigned materials such as films, periodicals etc.?

f.   Will make-up exams be allowed if I fail or miss a test?

g.   How much will each exam count toward passing the course?

Create a support system

Studying with others can be an advantage. They can share notes and help you review for tests.

Tip 2:   Develop a systematic method of studying.

If you do not develop this system, it will be difficult for you to be a successful test taker. Use the following rules to develop your system.

a. Learn the vocabulary of the course. Many students fail because they do not understand the vocabulary on exams.

Learn course vocabulary

b. Establish a study group or find a study partner from the class. Never go through a class without connecting with at least one other person in that class. You will have a built-in system for getting class notes you missed or for cross-checking information. In a study group, each person is responsible for units of information which they must share/teach to others in the group. This is good because when you have to teach information to someone else, you are more likely to learn it.

Find a study partner

c. Always scan your chapter before you read. Then read the introduction and summary. These two parts of the chapter should help you to zero in on what you need to know. Do not read your textbooks as if you are reading a suspense or romance novel. Read with the knowledge or understanding of what is on the next page. Make connections.

Read with a purpose

d. Always read with a question and/or objective in mind. Never read aimlessly. Take the information in like your exam is tomorrow. Know the objectives your instructor wants you to cover. For each unit covered, ask the instructor to share the *must know* items.

e. Always do a quick review before going to class.

f. Take good notes from lecture and/or assigned readings.

g. Immediately after each lecture/reading assignment, organize your information the way it should be remembered. Rewrite your notes if necessary. Preferably a rewrite should be done within 24 hours of the class/review.

The 24-Hour rule

h. Once material is organized, master it before you go back to class and take in new information. **Mapping** is a method that can help you organize material. See the example "Kinds of examinations" on pages page 175 and 176. Why don't you try making one for all of your courses.

Visual organization

Weekly Tests   i. It is a good practice to take weekly tests on new information. This will help in moving information from short-term to long term memory.

j. This is the time to develop mnemonics if necessary.

Flash Cards   k. Use flash cards. You may use these cards to review or test. You can take them places where you are not allowed to take books or tablets. A flash card is a 3"x5" card with the question on one side and the answer on the opposite side.

Test Specific Preparation—Part II

Tip 3:   Know How to "Get Ready."

Testing formats   a. Ask the instructor for the format of the test. Format refers to kinds of test such as **objective**—true/false, multiple choice, matching, fill in the blanks—and **subjective**—essay and short answer. Objective exams require precise information. Subjective exams usually are not as precise, but require you to support the data. The test formats are discussed later in this chapter. **Remember**, smart students test themselves before the instructor tests them. In preparing for an exam, they develop a test similar to what they think the instructor will give in class and take it under what they perceive as the same classroom testing conditions. Test preparation can be demanding but its the smart thing to do.

Exam questions   b. Ask the instructor what areas/chapters the test will cover. If you ask, some instructors will even tell you what questions will be on the test. Make sure you have all information from which the test questions will be taken.

Choosing a study mate   c. Study with a classmate whom you know does well in the course. It makes no sense to study with someone who is failing and cannot help you.

Test time   d. Establish the test date and how much time you will have to complete the test when it is given.

Pre-exam schedule   e. Set up a time schedule of pre-exam activities based on the date of the exam and your other life tasks (cooking, housekeeping, working, sharing). In preparing for the exam, you may have to sacrifice social, domestic and some sleeping time.

f. Develop test questions from lecture notes and assigned work.

g. Equally divide your objectives and/or anticipated test questions so that you learn and review a certain portion each day. Remember to consider the complexity of each question/objective. Learning a portion each day or over a period of time is called spaced learning. **Do not cram.** If you have not done so already, turn all objectives for the units or areas to be covered on the exam into questions. Write out your answers for all of them. If you are studying for a skill class such as mathematics, practice the processes until you are able to work them without guessing and/or repeatedly looking at your notes. You need to be able to clearly distinguish formulas and/or rules. **SKILL CLASSES REQUIRE A LOT OF PRACTICE.**

Spaced learning

h. When you sit down to study, always review all information previously studied before moving on to new study questions.

Practice! Practice!

Nicole is enrolled in PSY1000 and Professor Cooper has just announced a test for next Friday. He has given them thirty items to study and has indicated that the test will be a combination of objective and essay questions. Nicole works 12-16 hours a day on weekends. The only time she has for studying is during the week. So today she needs to make sure she has all of the notes and information from which the test will be drawn. Based on the test information, she has 4 days to prepare. She must also begin to try and determine which items may be asked objectively or in essay form.

Mon. 10 items	Tues. 10 items	Wed. 10 items	Thurs. Test/Rev	Fri. Test Day

Sample test schedule

Day One	Master the 1st 10 items/test yourself
Day Two	Review the 1st 10 items
	Master the 2nd 10 items/test yourself
Day Three	Review the 1st 20 items
	Master the 3rd 10 items/test yourself
Day Four	Retest yourself on all items/review
Day Five	Arrive early enough to relax/then take test

Note: You should use a realistic time table that takes into consideration your daily responsibilities and the difficulty of the material to be learned.

i. Ideally the day before exams should be for briefly reviewing materials and not trying to take in new data. Get a good night's rest and get to class early so you can have time to relax before the exam. **Do not** use this time to cram in new information. **Knowing the testable information is one of the greatest deterrents to test anxiety. Knowing the information really inspires confidence.**

Tip 4:   Know How to "Get Set."

a. Anticipate test questions and their format if you don't already have them.

b. Design a test on how you think the instructor is going to test you. Now test yourself with your notes and book closed just as they do in most class settings. Try to approximate the classroom environment. Set a timer or have someone call you when the time is up.

c. Grade your test through the eyes of your instructor. Review again if necessary.

Taking the Test—Part III

Tip 5:   Know How to "Go." You are in the classroom and have the test in your hand. You are now going to take the test.

Testing behavior

a. Read the directions. Make sure you understand them. Now is the time to ask for clarifications. Do not argue with the tester.

b. Scan the entire test before you begin.

c. Dump data as quickly as possible by lightly writing it on the back of your test or wherever it's feasible. This data refers to information that you are still holding in short term memory or that you have had difficulty retaining. **This may include acrostics and acronyms.**

d. Pay attention to qualifiers and absolute words.

e. Underline key words in directions and questions.

f. Now take a deep breath.

g. Develop a time strategy based on the number or questions and their point value.

h. Focus on your exam and do not let the behavior of other students distract you. For instance, some students panic when they see other students passing their papers in before them or busily writing when they are not.

Concentrate on test

i. If you are asked to do an essay, develop a brief outline first beginning with the main points.

j. When completing objective questions and you are using a scantron answer card, be sure to line up the questions and answers.

Scantron answer card

k. Read questions carefully and answer the easy ones first.

Answer easy ones first

l. When reading questions watch for grammatical agreements. This may help you in answering some of the items.

m. Look for answers in other questions.

n. Look for clue words.

o. Use the guessing or "bulling" rule if you don't know the answer. Bulling refers to coming as close as you possibly can with your answer. Be creative. Try to respond to all questions whether objective or essay. The only time you should leave a question unanswered is when there is a penalty and you are definitely unsure.

"Bulling"

p. Make time to look over your test before passing it in to the instructor.

Review after the Exam—Part IV

Tip 6:   Take a Deep Breath, Review and Modify If Necessary.

a. Reward and praise yourself if you did well on the test.

Analyze your strengths and weaknesses

b. Review answers to all questions you did not know and/or had to guess on before coming to the next class.

c. Analyze what you did right and what behaviors *you must modify* before the next test.

d. If you believe you did everything right and still did not pass the test, visit your instructor and let him/her give you analysis of the test items. It will help you to zero in on weak areas.

e. Immediately put into practice your new behaviors.

Take the time to study and prepare for tests. There is no magic. You must review and test yourself.

Kinds of Examinations

Type	Rules/Key words
Multiple-choice	1. Consist of stems, choices, distracters 2. Read the entire stem and choices. 3. Eliminate the distracters. 4. Select your answer from the remaining choices. 5. When more than one choice is correct, look for the choice "all of the above," or certain other designations. 6. When more than one choice is incorrect, look for the choice indicating some of the choices or none of the choices. 7. It saves time to answer the easy ones first. 8. There are few situations in which something is always or never true.
True-False	1. True/false tests are not the easiest. Don't be fooled. 2. Make sure you understand what is being asked. 3. You always have a 50-50 chance of getting the item correct. 4. For a true/false statement to be true, the entire statement must be true. 5. Conversely, if any part of the statement is false, the statement is false. 6. Beware of absolute words such as all, none, every, and always. 7. Statements containing qualifiers such as some, sometimes, most or often are frequently true.
Completion	1. You must know the correct information. 2. Sometimes clues are placed in the question. 3. Pay attention to words that precede the blank(s). 4. The grammatical structure of the sentence can help in determining what the instructor is looking for in an answer.
Matching	1. It is critical that you pay attention to directions. 2. Sometimes within the directions you will find that an answer may be used twice or only once. 3. If an item has only one correct answer, look for the more correct or complete answer. 4. As you collect responses (answers) cross them out so you don't waste time re-reading them again.

Type	Rules/Key words
Essay	1. Outline what you plan to cover. 2. Present your information in a clear, concise, neat and organized manner. This includes writing legibly. 3. Deal with the aspects of the topic the instructor has requested. 4. Pay attention to Key words since they tend to tell you how you are to deal with the topic. 5. Remember that most essay questions measure your depth of the knowledge of the subject and expect you to integrate and apply that knowledge. 6. State main points and use examples to support them. 7. Please refrain from excessive verbiage. Long answers do not equate to good grades.

Key Words

Discuss or explain	Examine in detail. Give facts, reasons, pros, cons.
Compare	Explain similarities and differences.
Contrast	Only explain the differences.
Critique/ Evaluate	Give your opinion on the good and bad aspects of the facts as presented.
Describe	Present a mental picture consisting of the characteristics and/or how something really is.
Define	Give the meaning (not your opinion).
Enumerate	State points and briefly explain one by one.
Illustrate	Use examples to explain.
Interpret	Use your own words to explain.
Identify	Be specific in listing/naming items in a category.
Outline	Describe the major facts/ideas that are relevant to a subject.
Prove/Justify	Use evidence or logic based on facts to support an argument/idea.
Relate	Show the connection between ideas and points.
Summarize	Present the main ideas usually in paragraph form.
Trace	Describe the progress or development of an event.

Practice Test: Practice your test-taking skills by taking the following timed quiz. Please follow directions. Before you begin, read all items carefully. Each question is worth ten points.

1.   Write your name here. _____

2.   Identify one concept that you know you have learned well in this

     course. _____

     _____

3.   Briefly explain it. _____

     _____

     _____

4.   When you compare and contrast, you are _____

     _____

5.   In a true/false question, there is always a _____ percent chance that

     the item is true.

6.   Multiple choice questions consist of the _____, _____

     and _____.

7.   In true/false items, in order for it to be _____ the entire

     statement must be true.

8.   In true/false questions, be aware of _____ .

9.   Once you receive your test in your hand, the first rule is to _____ .

10. Add the date beside your name in item number one and do not

     complete the other nine items.

How did you do? Question number one should be the only one answered with your name and today's date. The rest of the blanks should be empty. Subtract ten points for every one you attempted to answer.

Test Anxiety

Test Reactions

"I really studied for this test but when I got in there, I couldn't remember a thing." When I looked at the first question and couldn't answer it, I knew I was a goner." "I'm always afraid of exams because I never do well." "I'm so nervous, my hands sweat. I can't think and when I look around, everybody is working away but me."

Student Stress   *Elizabeth Stone*

Stress

Most of you have probably heard some of these statements before. They reflect how these persons are responding to test anxiety. Test anxiety is stress related to testing. Stress is defined as "the body's response to any demand made on it." The demand in this case is a combination of the test preparation and the test itself. On the one hand, you need to know that a little anxiety is good for everybody. It keeps us alert and on our toes. It keeps us from being too relaxed. If we are too relaxed, sometimes we take things for granted and don't prepare as we should. On the other hand, too much worry/anxiety keeps us from performing as we should. When one's

Memory loss

anxiety level is very high, we know that it interferes with your memory and as a result drives information from your mind that you know you are more than familiar with.

Anger, depression and a lack of confidence are other emotions that can also block memory. According to Atkinson and Longman, "test anxiety is a cycle in which self-doubt—even by a well-prepared student—causes the panic that results in a poor grade. The poor grade reinforces the feelings of self-doubt, which causes more panic and again self failure." When grades are tied to fear of loss and expectations of others, the anxiety level increases and once again causes one not to perform as is desired. Another factor to be considered is one's inability to concentrate on test preparation and the test once it is placed in your hands. You are not able to control your thoughts. Having counseled many students, I have found that the number one problem is really a lack of adequate preparation for the test. Most students simply are not test smart. An example of this would be the student who never tested himself/herself prior to the test under similar conditions as the instructor would do—closed book and timed. How do you respond to anxiety? Do you know the causes?

Role of emotions

Self-doubt

In the following exercise, think about what happens to you when you are anxious and what you think are the causes. For instance, a symptom might be a migraine headache. Working in a small group, complete the exercise below on your own paper and then share information. Once you have made your list of symptoms and causes, discuss ways to overcome them. See exercise 8.3.

Exercise 8.2

How Do You Know When You Have Test Anxiety?

Symptoms	Causes
1.	
2.	
3.	
4.	

Anxiety triggers

If you know that you have prepared well for the exam and you are failing, you may need to make an appointment with your school counselor. If you are one of those students who get ill at the thought of a test or upon walking into the test room, run to a professional counselor. You need help. If the reason you worry is because you have not studied, begin to do so now and eliminate your anxiety. Refer to the information on how to prepare for tests. However, the basic rule to rid oneself from normal test anxiety, is to learn to relax. Professionals try to teach you how to desensitize yourself to anxiety through a form of relaxation. The theory is that you cannot be tense and relaxed at the same time. Hence, if you know the information and you can get rid of thoughts that are blocking your memory and causing panic, you can get back control of your thoughts and memory. It sounds simple, right?

Test desensitization

During preparation and the exam itself, use deep breathing and muscle relaxing techniques to calm yourself. Close your eyes momentarily and breathe in and out focusing on something other than the exam. Focus by imagining a relaxing scene or situation. If that does not work, tense and relax your muscles while sitting in the chair with your feet flat on the floor and with your hands holding on to the sides of the chair. Tense and let go. Tense and let go. Do this four to five times while breathing in and out. To relax your hands, make tight fists and slowly release them. Do this a few times. In exercise 8.3, using your own paper, preferably the same one you used in exercise 8.2, tell us some of the things you do to calm your anxiety level.

Relaxation strategies

Exercise 8.3

Things I Do (or Should Do) to Get Rid of Test Anxiety

Anxiety reducers

1.
2.
3.
4.

? Journal Questions/Activities

1. Now that you have reviewed this chapter, share your current methods that you use in preparing for tests. What methods do you need to modify and/or change? Remember, its one thing to say and another to do.

2. Within the past week, how much time did you devote to test preparation? If you devoted any time at all to test preparation, what strategies did you use and why?

3. If you suffer from test anxiety, what strategies will you employ to reduce them. Discuss your plan of action.

Summary

This chapter has covered the reasons instructors give tests, the four parts of the test taking process and offered ideas on why we have test anxiety and how to cope with it.

Instructors give tests to assist the student in: mastering skills, learning course information, understanding how well they are progressing and whether they need to modify their method of studying for the course. Instructors also give tests to gauge how well they are teaching and getting the information over to students and whether they need to modify their method of delivery.

The four parts of the test taking process consist of: (1) *general preparation, (2) test specific preparation, (3) taking the test, and (4) reviewing immediately after the test*.

Test anxiety is stress related to testing, and this feeling is caused by physical and mental factors. Some are lack of confidence, inability to concentrate, fear of failure and the expectation of others. Being prepared for tests along with breathing and muscle relaxation exercises are some of the ways to cope with test anxiety.

Summary Exercise 8.4

The Art of Test Taking

Directions: Answer the following questions based on chapter information.

1. List three reasons why instructors give examinations and briefly explain your answers.

    a.

    b.

    c.

2. Identify the four parts of the test taking process and explain key elements of each.

    a.

    b.

    c.

    d.

3. Briefly define the following terms:

    a. Critique and/or evaluate

    b. Enumerate

    c. Discuss

    d. Illustrate

    e. Compare

4. Enumerate three reasons for test anxiety.

    a.

    b.

    c.

References

Atkinson, Rhonda Holt and Longman, Debbie G. *Getting Oriented*. West Publishing Company: New York, 1995.

Blerkom, Diana L. *College Study Skills: Becoming a Strategic Learner*. Wadsworth Publishing Company: Belmont, Calif., 1994.

Cirlin, Alan. *Simple Rules for Success in College*. Kendall/Hunt Publishing Company: Dubuque, Iowa, 1989.

Elliot, Chandler, H. *The Effective Student: A Constructive Method of Study*. Harper & Row, Publishers: New York, 1966.

Huff, Darrell. *Score: The Strategy of Taking Tests*. Appleton-Century-Crofts: New York, 1961.

Hawes, Gene R. and Hawes, Lynne Salop. *Hawes Guide to Successful Study Skills: How to Earn High Marks in York Courses and Tests*. New American Library: New York, 1981.

Herlin, Wayne R. and Albrecht, Laura J. *Study And Learning: The Development of Skill, Attitude And Style*. Kendall/Hunt Publishing Company: Dubuque, Iowa, 1990.

Kanar, Carol C. *The Confident Student*, 4th edition. Houghton Mifflin Company: New York, 2001.

Kessel-Turkel, Judi and Peterson, Franklynn. *Test Taking Strategies: How to Raise Your Score on All Types of Tests*. Contemporary Books, Inc.: Chicago, 1981.

Landsberger, Joe, site coordinator. Test Preparation and Taking Website, University of St. Thomas, St. Paul, Minnesota. 2004. `http://www.iss.stthomas.edu/tstprp1.htm`.

Maring, Gerald H., Burns, J. S. and Lee, Naomi P. *Mastering Study Skills: Making It Happen In College*. Kendall/Hunt Publishing Company: Dubuque, Iowa, 1988.

Starke, Mary C. *Strategies for College Success, Second Edition*. Prentice Hall: Englewood Cliffs, New Jersey, 1993.

Test Taking Strategies Website. Paul Treur, coordinator, University of Minnesota, Duluth. `http://www.d.umm.edu/student/loon/acad/strat/test_take.html`. 2004.

Chapter Nine

Career Planning
for the 21st Century

Ivan Stewart, M.A.   Leonard Koeth, Ed.D.   Sheryl M. Hartman, Ph.D.

Photo courtesy of Dr. Sheryl M. Hartman

Name _____   Date _____

Exercise 9.1

Career Choice Awareness Check

DIRECTIONS: Please place an "X" in the appropriate box.

	Yes	No	
1.	❐	❐	I should start planning for the future when I get out of school.
2.	❐	❐	Many people lack enough information to make a good career choice.
3.	❐	❐	My life goal is the same as my career goal.
4.	❐	❐	Many people are in careers that do not ideally suit their personalities.
5.	❐	❐	My abilities have little to do with my career choice.
6.	❐	❐	Abilities and aptitudes are the same.
7.	❐	❐	It will be easy to find many careers with work values such as security, high income, prestige, good future outlook, and a lot of leisure time.
8.	❐	❐	The three parts of the career planning puzzle are: (1) knowing yourself, (2) knowing about careers, (3) knowing how to get trained.
9.	❐	❐	I can learn about various careers on campus.
10.	❐	❐	Your career decision is only as good as the information on which you base it.

Introduction

This chapter will help you understand and solve the puzzles of career planning. Also, this chapter will help show you that planning your future will aid you in achieving many things you want from life.

Putting together the career puzzle

You need a plan

Exercise 9.2

My Dream Career

1. _____ Definite choice

2. _____ Leaning toward

3. _____ Undecided

It's so easy to dream about many things in life. But, if you don't plan you may find that your dream has turned into a nightmare! Undue optimism characterizes many vocational ambitions. Unfortunately this results in a series of adjustments to reality. These adjustments can create disillusionment and disappointment.

Even those fun things like taking a trip need planning. If you don't plan your trip, you may never arrive at your destination.

Just as failure to plan for details can spoil the trip, failure to plan your life and career goals can keep you from being all that you are capable of becoming and from achieving those things you want from life.

Now is the time for you to begin planning your future in terms of realistic life and career goals.

You must decide what you want from life, determine the occupation necessary to support your life style and decide if you are able and willing to get the education and training required for that career choice.

You must build your plan on a good foundation of information. The decisions you make are only as good as the information you base them on. Poor information—poor decision. Good information—good decision.

Students who focus on their career and select a major are more successful in college.

To solve the puzzle of career planning you need good information about three things:

1.   Know yourself (self-awareness)

2.   Career and job information

3.   Educational and training information

These three parts of the puzzle will fit together to make your career decision and career planning easy.

Know Yourself (Self-Awareness)

Who am I? You need to know a lot about yourself before you can decide what type of career may be best for you. It is important that you consider many types of information about you!

Ask yourself what you enjoy doing. You're the expert when it comes to knowing your interests. Do you enjoy working with people, data or things? Do you enjoy science, meeting people, selling, or music? Take the time to think about your likes and dislikes.

My Interest

Look at the occupational titles and carefully judge your degree of interest as you read the nature of the work for each occupation that catches your interest.

Visit the Online Occupational Outlook Handbook (OOH) at `http://www.bls.gov/oco/`.

Here you can search for specific information about an occupation.

You can get detailed information on the nature of work, and find out what these workers do. Included here would be information on the skills and responsibilities of workers in the profession, the use of technological advancements in the career field, and how job duties vary by industry or employer.

This site also provides information on working conditions. Working conditions include the workplace environment, the typical hours worked by employees, (e.g., evenings, weekends, a 40-hour work-week), the physical activities involved in the job, the special equipment you must be trained to use, and the extent of travel required.

Importantly, the OOH discusses typical earnings and how workers are compensated. The site reports on employment projections—the growth of the field or its decline. The geographical distribution of jobs is also reported.

The OOH site informs readers on how to train for a job, the training preferred by employers, the length of training, and advancement possibilities in the profession. Desirable skills, aptitudes, and personal characteristics are described, and the required certification or licensing for some fields are detailed.

You will also need to consider your aptitude for the occupation you have chosen. If you have the talents needed for the occupation you will find both the training and the job enjoyable.

My Abilities

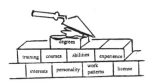

We all have strengths and weaknesses. You want your career choice to emphasize your strengths.

Remember that you first must get through the training program. Don't only look at the career at face value. It may look easy to you. Look carefully at the training program. What courses must I complete? What skills must I master? Be realistic! But remember that you have potential abilities (aptitudes) which may not yet have developed into skills.

Visit your Campus Career Center, where you can work with a counselor to assess your career interests, abilities and values.

The Educational Testing Services has a self-administered program—SIGI PLUS program, which is a self-directed, interactive system of career guidance. This guidance tool helps you to review all the major aspects of career decision making and this, or a similar product, is available in most schools and can be accessed over the Internet.

Complete Exercise 9.3.

My Work Values

What do you want from your occupation? The things you get from the job that you value are what we call work values. Values are what is really important to you in your daily life. Think about the following ten work values and decide which are most important to you:

1.   How much prestige do you want from the job?

2.   Do you want a high degree of security on the job?

3.   Is earning a high income important to you?

4.   How much leisure time do you expect?

5.   Is being in a leadership position important to you?

6.   What about independence on the job?

7.   Do you want variety?

8.   How about employment outlook? Do you want a career that is growing?

9.   Are you one that likes helping others?

10.   Do you want to work in the field in which you are most interested?

All these work values may be desirable, however, no one occupation has all these desirable values. You must do some value clarification and decide which you can do without and which you need. You then must search for the career that best gives you those values. Of the ten values, you just looked at, which three do you feel are most important. Also find two that you would like to have but are not a must. Now complete Exercise 9.4.

Name _____   Date _____

Exercise 9.3

Identify Your Achievements and Aptitudes

1. For the last 2-3 years, list the subject in which you got the best grades:

    _____

    _____

    _____

2. List the subjects you like best:

    _____

    _____

    _____

3. What are the subjects that you like best and in which you also receive the best grades?

    _____

    _____

    _____

4. From your past experience, estimate your possible aptitude level for each of these work aptitudes.

	High	Medium	Low
a. Business Aptitude	____	____	____
b. Clerical Aptitude	____	____	____
c. Logical Reasoning	____	____	____
d. Mechanical Reasoning	____	____	____
e. Mathematical Aptitude	____	____	____
f. Social Skills	____	____	____

Exercise 9.4

My Work Values

1. List Your top 3 work values

    _____

    _____

    _____

2. List 2 or more values that are not a must but would be nice to have.

    _____

    _____

Another part of the puzzle deals with your preferred work patterns. What are your thoughts on working weekends, overtime, shift work or irregular hours? Do you mind seasonal work or traveling? What about sitting for long periods? Do you mind working outside or in extreme cold or heat? What about noise, odors or dampness? These are only some of the work patterns you may encounter on the job.

My Work Patterns

You may have considered some of the previous factors in helping make your career choice. But you never gave your personality a thought. It is found that personality is an important factor related to job satisfaction.

My Personality

People many times adjust themselves or become something they are not, hoping to achieve some end. You may go to a job interview and pretend you love to work with people. After all the job is in sales and pays a high income. You get the job and soon you are being stressed due to the high amount of interaction with people. Actually you like working with things better than people. You have play acted yourself into being one of those unsatisfied with their career choice because you tried to adjust your personality.

Personality is a complex subject. However, over a period of many years researchers have accumulated a considerable amount of data that relates to personality and career choice. By using information about your personality you can be matched with careers in which you can be comfortable being yourself.

You may want to read more about personality or ask your instructor how you can take a personality assessment such as the Myers-Briggs Type Indicator which is very popular with career counselors.

Career and Job Information

Now that you know yourself you will need to know about careers. Then you can put parts of the puzzle together. You must find out about different occupations and see where you fit in. You will be identifying occupations that match your interest, abilities, work patterns and personality.

Complete Exercise 9.5 for two careers.

Exercise 9.5

Career Write-Up

Sources used _____

Job title _____

Nature of work

Working Conditions

Employment (Where are the jobs?)

Training, Qualifications, and Advancement

Job Outlook

Earnings

Related Occupations

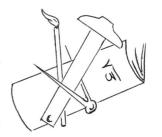

Educational and Training Information

The final piece of the puzzle is the training program. You must find the best one for you. Again you must find information about a number of things to make a good match.

Now that you know your major you must see which program suits you best. What length of program do you want? Do you want a certificate or occupational program that takes a year or less to complete? Are you interested in a two-year technical program or a two year Associate in Science Degree? Does your career choice require a license, certification or internship? Can you meet admission standards? Is the program limited access (you compete with others to get one of the limited slots in the program)? Does your career choice require you to get a bachelor's, master's or doctorate degree? What are the required courses you must complete? What will it cost? Where is the training? Does your academic ability and interests match that of the program? Does the program match your major goals?

To search out this information you can visit your career center or library to find admission requirements. You can talk to advisors and talk to people who have gone through the program. Send for flyers and catalogs. Visit the school if you can.

Complete Exercise 9.6.

University Transfer and Scholarships

Once you have decided on a career and choice of major you must decide if you are going on to a university. If the answer is yes, you must choose the right university for you. You must identify what you want from a university and then find those universities that match what you want. You will want to compare universities by getting information from various sources. You can find information from sources such as the *Blue Books*, *Lovejoy's*, *The College Handbook* and others. Research scholarship information.

Name _____   Date _____

Training Decision Chart

Rate each category listed below using the following scale:

2-very acceptable   1-acceptable   0-not acceptable

CAREER CHOICE	SCHOOL	COURSE OR TRAINING REQUIRED	LICENSE, CERTIFICATION, INTERNSHIP, OR DEGREES	COST	LOCATION	TOTAL
_____	_____	_____	_____	_____	_____	_____
_____	_____	_____	_____	_____	_____	_____
_____	_____	_____	_____	_____	_____	_____
_____	_____	_____	_____	_____	_____	_____
_____	_____	_____	_____	_____	_____	_____

Using the rating, choose the training programs that seem like the best choices.

1. _____   2. _____

There are many websites which provide detailed information on specific schools, online applications, and student financial aid information. To find further college and university information you may wish to visit these websites:

```
http://www.usnews.com/usnews/edu/college/cohome.htm
http://www.college-scholarships.com/
http://www.collegeview.com/
http://www.petersons.com/
```

Complete Exercise 9.7.

Exercise 9.7

Matching Sheet

Matching Personal and Occupational Information

Directions: Write the titles of two occupations you've explored (A and B) at the top of the columns. Read each statement under self-awareness. If the statement is correct for either or both occupations put a check on the line under that occupation. If neither occupation matches your personal traits put a check on the line under neither.

SELF-AWARENESS	OCCUPATION A	OCCUPATION B	NEITHER
Interests My personal interests seem to match the interest areas related to this occupation.	_____	_____	_____
Abilities I have the abilities that this occupation requires. My plans for furture education or training are at least equal to the amount and type this occupation requires.	_____	_____	_____
Values (What's important to me) The employment outlook for this occupation gives me the job security I need.	_____	_____	_____
The chances for advancement in this occupation are good enough to satisfy me.	_____	_____	_____
The pay range in this occupation is high enough to satisfy me.	_____	_____	_____
Patterns My work patterns fit in with the working conditions typical of this occupation.	_____	_____	_____
Personality My personality seems to be right for this occupation and I can be myself on the job.	_____	_____	_____

[?] Journal Questions/Activities

1. What is your career goal?

2. What have you done to propel yourself toward achieving your career goal?

3. What will you need to do in the future to make this goal a reality?

4. What values are most important to you in your career choice?

5. Identify at least two things you learned about yourself or your career choice from this chapter.

6. Did this chapter give you what you expected? If not, what more do you need?

Summary

This chapter helped you put together the very difficult puzzle of career choice. You have researched each piece of the puzzle and put them together. You now must outline your plan and follow it to be successful.

You found out about yourself by getting information about your interests, abilities, work values, work patterns and personality. Then you researched career and educational information to match to your personal characteristics. You have used a planned and organized way of making your career decision.

Name _____   Date _____

Summary Exercise 9.8

DIRECTIONS: Please place an "X" in the appropriate box.

	Yes	No	
1.	❐	❐	I should start planning for my future right now and not wait until I am out of school.
2.	❐	❐	Most people who chose their career usually base their choice on good information.
3.	❐	❐	My career goal is a part of my life goal.
4.	❐	❐	Many people are in careers that ideally suit their personalities.
5.	❐	❐	My abilities have an important bearing on my career choice.
6.	❐	❐	Aptitudes are potential abilities.
7.	❐	❐	There are few careers with all the desirable work values such as security, high income, prestige, good future outlook, and a lot of leisure time.
8.	❐	❐	The three parts of the career planning puzzle are: (1) Luck, (2) Chance, (3) Taking whatever is available.
9.	❐	❐	There are no sources on campus to get information on careers, universities or scholarships. You must GO off campus.
10.	❐	❐	You can make a good career decision without much information or planning.

References

Lock, Robert. *Student Activities For Taking Charge of Your Direction*. 2nd ed. Pacific Grove, CA: Brooks Cole, 1992.

Sukiennik, Diane, et al. *The Career Fitness Program*. Scottsdale, AZ: Gorsuch Scarisbrick, 1999.

U.S. Department of Labor, Bureau of Labor Statistics. *Occupational Outlook Handbook*. Washington, D.C. 2003.

Chapter Ten

Academic Regulations

Sheryl M. Hartman, Ph.D.

Exercise 10.1

Academic Regulations Awareness Check

DIRECTIONS: Please place an "X" in the appropriate box.

	Yes	No	
1.	❏	❏	The *Degree Audit* includes specific courses required for each general education and program area.
2.	❏	❏	Students may now "degree-shop" online, to determine immediately how their completed courses satisfy a variety of majors.
3.	❏	❏	SOAP stands for Students on Academic Probation.
4.	❏	❏	A student with a 1.75 GPA is on probation.
5.	❏	❏	All students must take the CLAST to graduate.
6.	❏	❏	To drop a class you must see your instructor.
7.	❏	❏	Withdrawals count when calculating a grade point average.
8.	❏	❏	You must first register for a class in order to audit it.
9.	❏	❏	A co-requisite is a course taken at the same time as another course.
10.	❏	❏	All higher education institutions use a 4.0 grading system.

Introduction[1]

These regulations, procedures and academic terms are published to assist students by providing information that is essential for planning, pursuing, and understanding their academic programs, leading to the achievement of educational goals.

Academic Regulations

Each college has guidelines to assist students in progressing in a steady path toward their academic goal. Learning new terms used in the academic environment and understanding their meaning can impact your comfort in the college atmosphere. This is turn impacts the duration of time it takes to complete academic programs, the quality of your academic work, your readiness for academic courses, and the transferability of your educational activities.

Most importantly, you need to be in control of your college experience. Not your advisor, not your parent, not your significant other. YOU! You need to understand the important decisions you are making in the academic environment.

Academic Definitions

ADMISSION—Most college admission forms may be viewed and printed from any college website. You may also take a virtual tour of many campuses. It is traditional to require information on residency status, high school graduation, and past academic credentials during the application process. Oftentimes, an institution will allow a student to enroll for one academic semester during what is termed a "grace period" while this information is being requested and submitted.

REGISTRATION—Registration involves a student enrolling in classes. There are several methods of registration: face-to-face with an academic advisor, touchtone telephone registration, and Web registration.

During the registration process, an ID picture is taken, and a smartcard may be disbursed. The Smartcard functions much like a credit card, with some institutions allowing the student to use this card in their College Bookstore—accessing their money from their financial aid package.

Important Terms

---

[1]   Academic regulations presented here apply to students seeking degrees from Miami Dade College only. Partner institutions may have various academic policies and regulations that differ from MDC about which you may wish to research.

Degree Terms

ASSOCIATE IN ARTS DEGREE—The Associate of Arts degree is awarded to students completing the requirements of the academic transfer program with a minimum of 60 semester hours[2] including 24 hours of general education courses. This is the degree that would be most appropriate for the majority of transfer students because it parallels the work done in the first two years of a Bachelor of Arts degree at a four-year institution. This degree is designed for the student wishing to transfer to upper division universities. Many institutions have articulation agreements with four year institutions that allow you to enter these schools as a junior when you have completed the Associate in Arts Degree.[3]

ASSOCIATE IN SCIENCE DEGREE—The Associate of Science degree is awarded to students completing the requirements of specifically identified programs. These areas of study are designed primarily to prepare students for immediate employment. Credits earned for many courses in these programs (e.g.: Occupational Education Allied Health Programs) may be applied to upper division colleges if the student chooses to move on to a Bachelor of Science Degree Program.

ADDITIONAL TERMS

Audit

An auditor is a student who enrolls in a course, attends class, pays full fees, but does not receive a grade or credit for the course. Audit status is granted on a space available basis and with institutional consent. Audit status may not be changed to credit status. Audited courses are reported on the student's academic record with non-punitive notation.

COLLEGE LEVEL ACADEMIC SKILLS TEST (CLAST)—The CLAST will measure competencies in reading comprehension, writing and computational skills, including Arithmetic, Algebra, Geometry Measurements, Logic Sets, Probability and Statistics.

In Florida, all students must show competence in English language, reading, and mathematics skills before obtaining an Associate in Arts

---

[2]   Miami Dade College requirements

[3]   This is a partial list of the institutions with whom Miami Dade College has developed articulation agreements: All Florida State Institutions
Polytechnic University of the Americas
University of Miami, Coral Gables
Barry University, Miami Shores
Massachusetts Institute of Technology
California State University at Dominquez Hills (CSUDH)
Drexel University, Philadelphia
Brown University, Rhode Island

degree. To demonstrate this competence, every student must satisfy one of the following alternatives:

1. Receive passing scores on all four subtests of the College Level Academic Skills Test.

2. Earn a Grade Point Average of at least 2.5 in specified courses taken in college.

3. Achieve satisfactory scores on the SAT or ACT.

Students are eligible to write the CLAST after they have earned 18 college-level credits, exclusive of college preparatory and ESL/ENS credits, and successfully completed English 1101 with a grade of "C" or above for the communications subtests and a college-level mathematics course with "C" or above for the mathematics subtests.

COLLEGE PREPARATORY CLASSES—All regularly admitted first-year students who wish to enter degree programs must take a college placement test which assesses their current skills in English, reading, and math. Students are placed into college preparatory courses based on their need for remedial word as indicated by this assessment. This assessment can alternatively be met by submitting test scores from the Scholastic Achievement Test. You can practice for this with the SAT Question of the day at http://www.collegeboard.com/apps/qotd/question.

COMPUTER COURTYARD—Computer resources on campus. Most campuses have several computer labs, including some with the latest word processing programs, additional labs with tutorials in specific subjects, computer animation tools, film editing tools, data base management software and business use software.

CO-REQUISITE—Some courses are required to be taken together. An example of this would be a chemistry lecture course and a chemistry laboratory course. When a course has a co-requisite, this indicates that both courses must be taken at the same time.

CREDIT—A college credit is a numerical unit received for completing a course.

DEGREE AUDIT REPORT AGIS—This report is an advisement tool students can access which will provide information on how they are doing in meeting their degree. The Degree Audit tool can be manipulated by the students to review different academic scenarios. The Degree Audit can be used to help the student build their schedule and register. This dynamic tool may also be used to help with the proper sequencing of courses.

DROP—To drop is to withdraw from a class. When a student drops a class prior to the 100% refund date, this class does not show on their academic record. Classes dropped during the academic term show as a "W" on the Academic Record.

FULL-TIME STUDENT—This is a student enrolled in at least 12 credits in a major term or at least six credits in a minor term. Courses and credits enrolled for audit do not count in the computation of full-time or part-time enrollment status.

GRADE POINT AVERAGE (GPA)—Each letter grade has a point value. A grade of "A" is interpreted as excellent and has a point value of 4, a "B" is interpreted as good and has a point value of 3, a "C" is average and has a point value of 2, a "D" is poor and has a point value of 1, and both an "F" (failure) and "U" (unsatisfactory) have a point value of 0. The figure obtained when dividing the number of credits attempted into the number of grade points earned is the GPA.

GRADE POINTS—Also known as quality points, these are the numerical values assigned to grades earned. The grade points for each course are determined by multiplying the number of points a grade is worth times the number of credits the course carries. For example, an "A" in ENC 1101 (3 credit course) is worth 12 grade points, while a "B" in this same course would be worth 9 grade points.

INCOMPLETE—When a student has requested and received an incomplete grade, he/she must complete a written and signed contract with their instructor regarding the work that needs to be completed and the time period they have for this effort. Incomplete grades do not remain on a transcript indefinitely; after 12 months they are automatically programmed to be changed to a grade of "F."

MAJOR TERM—Major terms consist of approximately 16 weeks.

MINOR TERM—Minor terms consist of a six-week period. Some courses are scheduled for a 12-week period, combining two minor terms.

PRE-SELECT—This is a term used to identify initial student interested in highly competitive programs of study. These programs, such as medical sonography and physical therapy, have limited openings and students are not guaranteed acceptance. A separate application following preliminary course work is required.

PREREQUISITE—This is a requirement which must be met before a certain course may be taken.

STANDARDS OF ACADEMIC PROGRESS (SOAP)—This is a system of academic standards based on a student's GPA and the number of courses attempted, completed, and withdrawn from. Students who experience academic difficulty are alerted through the use of these Standards.

STAR (Student Telephone-Assisted Registration)—A computerized telephone registration system which allows students to register from a touch-tone phone.

Substitution—The replacement of a course required in a program with another course specified by the school/division or department which established the requirement.

Transcript—A certified copy of a student's permanent academic record on file listing each course and the final grade received.

Virtual College—Virtual College is a way of extending education to those students who wish to take courses via the Internet. If you have a work schedule or a personal situation that prevents you from attending regular classes, and if you have ready access to a computer and the Internet, you may complete degree courses online. All courses in the Virtual College have Internet support (online learning forums, chat, and e-mail) that will help the student to communicate with classmates and with their professors.

Vocational Credit Certificate—Vocational Credit Certificate Programs prepare students to enter a specific career or vocation. To complete a program students must demonstrate that they have mastered specific job-related performance requirements as well as communication and computation competencies.

Withdrawal—This is the act of officially leaving a class or the College.

Standards of Academic Progress (SOAP)

The main purpose for the Standards of Academic Progress is to establish a formal process through which the administration and faculty can identify and provide assistance to students who experience academic difficulty. When academic progress has not been satisfactory, the Standards of Academic Progress require students to limit the number of credits for which they register.

The Standards are not intended to discourage or penalize students who are sincerely trying to make good use of the college's instructional services. Rather, they reflect the commitment of faculty and administration to provide students with as much assistance as possible to insure success in achieving their educational goals. Students have available to them a variety of means to remedy their academic weaknesses. Special academic assistance will be provided by the college to those students. The overall objective of the Standards is to inform students about their academic performance, so that they can make corrections as rapidly as possible.

Definitions:
Standards of Academic Progress Results

Restrictions placed on SOAP students

Academic warning. This limits student enrollment to 12 credits in the major terms and 6 credits in the minor terms. This includes 3 or more credits that are part of a prescribed program of intervention which attempts to assist the student. Academic warning intervention may include college preparatory courses, a study skills course, career counseling or a combination of all three.

Academic probation. This limits student enrollment to 9 credits in the major terms and 3 credits in the minor terms. This limitation includes 3 or more credits of prescribed intervention courses. Students remain on academic probation until they maintain a 2.0 overall grade point average and earn credit in at least half the credits for which they register.

Academic suspension requires students to discontinue enrollment for one major term.

Note: A student completing a successful appeal of their academic suspension may be permitted to register. The student will be required to earn at least half of their credits each term and maintain a 2.0 term grade point average. The student will be on academic probation.

Academic dismissal represents a separation of students from the college for a twelve-month period. Academic dismissal occurs after students fail to meet the minimum requirements during probation after suspension. If, after being readmitted following suspension, the student fails to meet minimum standards, which are maintaining a 2.0 term grade point average and earning credit in at least half the courses for which he/she is registered, he/she will be separated from the college.

Students are eligible to apply for readmission to the college after the dismissal period. Admission will be on a petition basis. In order for readmission to be approved, the petition must present evidence of some change in the student's circumstances.

The Degree Audit System

The Degree Audit report is a dynamic interactive tool. Data from this report will help students to develop a time plan for their college enrollment period. It will also be used to assist in student advisement and show student progress toward meeting graduation requirements.

The Degree Audit includes specific courses required for each general education and program area; the student's standing using the Standards of Academic Progress (SOAP); grade point average; and posting of automatic graduation remarks. The Degree Audit report will connect students to the Internet, allowing them to compare academic programs and to immediately determine whether their completed courses may satisfy requirements for several majors.

Withdrawing from a Class/College

Process for dropping classes

If you find you cannot complete a class for any reason, you must withdraw from the class in order to avoid receiving a failing grade. Your instructor does not know why you stopped coming to class unless you process the drop card in the registrar's office. As a courtesy, you should inform your instructor that you are dropping the class. If you find that you cannot complete any of your classes in a term, you must withdraw completely from the College by completing the drop card in the registrar's office. You will receive a "W" for any class you drop if you process the drop before the deadline. Drop dates are published in the college catalog, the registration handbook, and the academic calendar each term. You are responsible for knowing these dates.

If you drop a class before the 100% refund deadline at the beginning of the term, you will receive a check for the full amount of the class and the course will not appear on your schedule for that term. No grade will be assigned for that class.

If you drop a class after the 100% refund deadline, you will receive a "W" as your grade, but no refund will be given. A "W" does not calculate in your grade point average, but an "F" does. Therefore, you must be careful to observe the drop deadlines.

Grade Point Average

Grade Point Averages (GPA) at Miami-Dade are calculated on a 4.0 scale. Quality points are assigned for each grade as follows.

A = 4 points per credit	U = 0 points per credit
B = 3 points per credit	W = no calculation
C = 2 points per credit	P = no calculation
D = 1 point per credit	S = no calculation
F = 0 points per credit	X = no calculation

A 2.00 GPA is required for graduation. Grade point averages are not rounded-out so it is necessary to maintain at least 2.00 to qualify for graduation.

Instructions for calculating grade point average

Determine how much each of your grades is worth. For example, if you received an "A" in your three-credit English class, you would calculate the following:

English   3 credits A

Multiply 3 (credits) times 4 (each credit of "A") for a total of 12 quality points.

Use the same procedure for a four-credit math class with a "C" grade.

Multiply 4 (credits) times 2 (each credit of "C") for a total of 8.

Note the following sample of GPA calculation:

English	3 credits	A	3 x 4 = 12
Social Sci	3 credits	C	3 x 2 = 6
Chemistry	4 credits	B	4 x 3 = 12
Phys. Ed.	2 credits	F	2 x 0 = 0
SLS 1101	1 credit	B	1 x 3 = 3
Math	(3 credits)	W	Does not count
Total	13 credits attempted		33 quality points

Now divide the total number of credits attempted into the total number of quality points.

$$
\begin{array}{r}
2.53 \\
13\ \overline{)\ 33.00} \\
\underline{26.} \\
70 \\
\underline{65} \\
50 \\
\underline{39} \\
11
\end{array}
$$

The grade point average is 2.53

Exercise 10.2

Calculate the following grade point average:

English	3 credits	B
Soc. Sci	3 credits	D
Physics	4 credits	C
Algebra	3 credits	F
Phys. Ed.	2 credits	A
SLS 1101	1 credit	A

Exercise 10.3

Now calculate the same grade point average but change the algebra grade to "W" instead of "F." Notice the difference a withdrawal can make in your grade point average.

? **Journal Questions/Activities**

1.   What did you expect to learn from this chapter?

2.   Which regulations will be more difficult to complete?

3.   What information would you recommend be added to college websites?

Summary

You have taken several very important steps toward achieving your academic goals. You are in an educational environment, you are becoming proficient with academic terminology, and you have begun to recognize your role and your responsibilities in executing your educational plan. Being willing to enter this new and unfamiliar setting, and learning the new skills needed to master it, demonstrate your courage and determination. These are strengths that will serve you well.

Name _____   Date _____

Summary Exercise 10.4

1. What is a full-time student?

_____

2. What is a credit?

_____

3. What is the CLAST?

_____

4. What is a prerequisite?

_____

5. What is a transcript?

_____

6. What is SOAP?

_____

7. What is Virtual College?

_____

8. What does the term "pre-select" mean?

_____

It is very interesting to explain college websites. You might want to take a look at the following site, http://www.mdc.edu. If you look down the left-hand frame, you can link to the academic calendar, news articles of interest, reports of college graduates and more. Most colleges now have a very detailed presentation on their websites.

Add some of your favorite sites here.

_____

_____

_____

_____

_____

_____

_____

_____

_____

_____

_____

Chapter Eleven

Diversity in Education

Carol Cooper, Ed.D.

Exercise 11.1

Diversity Awareness Check

DIRECTIONS: Please consider the following statements. For a statement you believe to be true, place an "X" in the appropriate box. If you believe a statement is false, place an "X" in the appropriate box. Discuss your answers.

	True	False	
1.	❐	❐	I am aware of the many cultures that are represented in my community.
2.	❐	❐	Cultural differences do not affect my choice of friends.
3.	❐	❐	My friends do not influence my choice of friends.
4.	❐	❐	People differences will not affect where I plan to work once I get my degree.
5.	❐	❐	It is okay to be prejudiced.
6.	❐	❐	Anti-semitism is a form of racism.
7.	❐	❐	I believe discrimination occurs when a person is kept from entering an activity because of the group of people to which he belongs.
8.	❐	❐	Native Americans are the white European settlers that landed in 1617.
9.	❐	❐	Language plays a role in discrimination.
10.	❐	❐	One's skin color does not affect "preference."
11.	❐	❐	Success does not depend on whether you are red, yellow, black or white.
12.	❐	❐	I honor the "Golden Rule" when dealing with people.
13.	❐	❐	All races are created equal.
14.	❐	❐	The September 11, 2001 attack (when planes flew into the World Trade Center in New York, among other targets) was incited by American values.

Introduction

As you embark on your educational journey, you will begin to experience more diversity than you have ever had in your life. The United States adds an immigrant every twenty-five seconds and they come from around the world. The shift from a western European society is very evident in the population and is also mirrored in college life. Colleges also accept students from around the world who bring their own traditions, values, beliefs and religions. Can you imagine the lifetime of experiences which you and all of these students bring to the academic environment? To be sure, it is stimulating. However, there may be some problems. Stuart expresses it this way:

> *You may experience emotions such as anxiety, anger, embarrassment, or guilt as you begin to learn about and appreciate perspectives that differ from your own. It is worth persevering in the effort because you will broaden your experience and learn skills that you will need as you enter the more diverse world of the twenty-first century.*[1]

Fortunately for you, the college makeup reflects the community and the nation-at-large in terms of population and trends in the multicultural environment. This allows you an opportunity to enhance and modify your human relation skills. On the other hand, it will also force us to look at the whole spectrum of diversity issues such as discrimination and harrassment based on race and gender. In this chapter we will cover the population makeup of the United States and the county in which you live. It is important for you to prepare yourself for the future by understanding where you will work and live. It is also important for you to understand how diversity can affect your education. In both instances, you must be able to communicate effectively. In order to achieve this, we will look at the American value system, common habits and beliefs that tend to cause problems in our diverse and changing society. We will also provide insights into societal conflicts and what you can do to learn how to be sensitive to differences.

United States demography

The population of the United States is over 292 million people. Of this number, whites comprised 83 percent; blacks, 12.9 percent; Asians and Pacific Islanders, 4.1 percent; and American Indians, Eskimos and Aleuts, .9 percent.

African Americans

The African Americans constitute one of the largest minorities in the U.S. They come from areas such as Africa, the Caribbean, and South and Central America.

Asian/Pacific Islanders

The Asian/Pacific Islanders have almost doubled their population. This group is composed of people coming from places such as China, Japan, Philippines, Korea, Asia, Vietnam, Cambodia, Laos and Thailand.

[1.]   Starke, Mary C. *Survival Skills for College*. New Jersey: Prentice Hall, 1993, p. 253.

**Alaska Natives
American Indians**

Hispanics

The Eskimos, the Aleuts and some of the Native Americans come from Alaska. The rest of the American Indians comprise about 300 tribes located throughout the U.S.

The Hispanics compose the fastest-growing minority in the U.S. They come from Spanish-speaking areas such as Mexico, Cuba, Puerto Rico and Central America. The Puerto Ricans are U.S. citizens. Their population estimate as of 2000 was that they comprised approximately 13 percent of the population.

It is projected that in the year 2050, these minorities will comprise 47 percent of the U.S. population.

The students, staff and faculty whom you will encounter come from different backgrounds, ethnicities, races and socioeconomic levels with many different belief systems. This richness of diversity will only enhance your educational experience.

Keyser and Keyser believe

that valuing diversity will help us solve problems. Other authors have reported that diverse groups are able to solve problems faster and more effectively than homogeneous groups. They further stated that children attending integrated schools acquire greater social skills, achieve higher levels of personal growth . . . and are more likely to be able to adapt to a wider range of environments.[1]

How will you cope in such a diverse environment while dealing with your education? Can you tolerate other cultures? Can you see beyond your own point of view? How did you fare on the diversity check?

In today's society, everyone must learn to be sensitive and intelligent when dealing with others who are different. The ability to adapt to changing environments is vitally important to future success. Everyone will not need to like everyone else, but everyone must make an attempt to understand everyone else if people are going to work and live together. Have you taken the time to learn the other cultures in your environment? Before you look at your community, look at some of the terms that tend to cause problems in diverse environments.

The United States of America was attacked on September 11, 2001 by terrorists who used airplanes loaded with passengers to crash into the World Trade Center and the Pentagon. Another airplane crashed before it could reach its intended target. This terrorism demonstrates why it is important to understand and value diversity.

1. Keyser, John and Marilyn. "From Affirmative Action to Affirming Diversity." *Community, Technical, and Junior College Journal*, vol. 61, No. 3, Dec./Jan., 1990/91, p. 7.

Terms Dealing with Diversity

American Values

Just as individual behavior is guided by one's value system, so is American society guided by a "Value system." It is based on the Judeo-Christian religious ethic and heritage. Basically what this says is that we live in a society predicated on "freedom, democracy and fairness." Within freedom, democracy, and fairness, individuals are to value responsibility, productivity, community, family and work. Christians comprise approximately 84 percent of the population. Our legal system is still based on the Jewish/Christian Bible. Our coins read "In God we trust."

Diversity

Diversity refers to differences in society. Differences include male and female; black and white; Hispanic, African, Asian, Native American; young and old; able and disabled; socioeconomic status; sexual and religious preferences.

Culture refers to learned attitudes, beliefs, and behaviors that characterize a particular population. It includes such features as how we dress, body language, what we eat, music we listen to, how we make products and sometimes how we see ourselves versus the rest of society. We are all molded and shaped by culture.

In a complex society such as America, there are usually minor variations in lifestyle and behavior from group to group. These variations are often associated with difference in ethnicity and social class, with social class as a much more significant factor than ethnicity.

Ethnicity

An ethnic group is a group whose members **identify** with one another because they share a common culture. Race, religion, and ethnicity are often confused with each other with unfortunate results.

In South Florida, for example, blacks are often thought of as an ethnic group, but American-born blacks are different ethnically from Bahamian-born, Jamaican-born, or Haitian-born blacks who all have distinct identities.

Likewise, people often speak of Hispanics without realizing that many Cubans, Puerto Ricans, Nicaraguans, and Argentineans would resent being lumped together in one ethnic group just because they all happen to speak Spanish.

Jews are regarded as an ethnic group in the United States because of their common religious heritage, but in Israel, there are very distinct lines drawn among ethnic Jews from Russia, Germany, Yemen, Morocco, etc. In the United States, religion provides Jews with a source of ethnic cohesion, but the same is not true of most other religions in America. Both the Catholic and Protestant branches of Christianity contain dozens of ethnic nationalities.

American Indians are racially distinct from white Anglos, but each major tribal group, i.e. Navahos, Hopi, Sioux, etc., is composed of different cultural traditions.

In summary, there are numerous ethnic identities in America. Only in isolated cases is an ethnic group represented exclusively by an entire race or religion.

Race Race has traditionally referred to a group of people who share inherited characteristics of outward appearance such as skin color, hair texture, stature, and facial features. Race should not be confused with culture or ethnicity, or religion, or kinship groups.

Racism Racism couples the false assumption that race determines psychological and cultural traits with the belief that one race is superior to another. Based on their belief in the inferiority of certain groups, racists justify discriminating against, segregating, and/or scapegoating these groups. Racists, in the name of protecting their race from contamination, justify the domination and sometimes even the destruction of those races they consider inferior.

Prejudice Prejudice is a set of rigid and unfavorable attitudes which are formed in disregard of facts toward a particular group or groups. It is an unsupported judgment usually accompanied by disapproval.

Discrimination Discrimination is differential treatment based on unfair categorization. It is denial of justice prompted by prejudice. When people act on their prejudices, they engage in discrimination. Discrimination often involves keeping people out of activities or places because of the group to which they belong.

Scapegoating Scapegoating refers to the deliberate policy of blaming an individual or group when the fault actually lies elsewhere. It means blaming a group or individual for things they really did not do. Prejudicial attitudes and discriminatory acts lead to scapegoating. Members of the disliked groups are denied employment, housing, political rights, and social privileges. Scapegoating can lead to verbal and physical violence as well as death.

Anti-semitism Anti-semitism is prejudice or discrimination against Jews based on negative perceptions of their religious beliefs and/or on negative group stereotypes. Anti-semitism can also be a form of racism as when Nazis and others consider Jews an inferior "race."

Stereotyping Stereotyping is the process of lumping perceptions into broad categories and processing these perceptions on the basis of what the categories are like rather than on the unique characteristics of each individual or object. This process extends into all aspects of life. Because your brain is constantly inundated with what is going on around you, you have to find a way to process this mass of information. You "tune out" or ignore perhaps 99% of all incoming stimuli and attempt to organize the remaining 1% in a manner that is meaningful to you.

People tend to stereotype other people and objects in a way that is meaningful to them. They classify people according to their status and group memberships. Thus, they put teachers, students, old people, police, males, black males, union members, politicians, and children into stereotyped categories. They then prepare to deal with them in a manner considered appropriate to them based on previous experience and what they have heard.

Some stereotyping is normal and universally-used as a cognitive tool. What about the bad kind of stereotyping associated with racism, sexism, and various kinds of discrimination? Where can the line be drawn between one type and another? It needs to be understood that the process of stereotyping is the same, regardless of the object of the stereotype or consequences that flow from it.

Stereotyping becomes socially-undesirable only when people use it to limit their interactions with others or when it inhibits the ability of people to form and maintain significant and constructive social relationships. Therefore, one needs to exercise a measure of insight and self-awareness when relating to others. One must constantly ask oneself, "Is my reaction to this individual based on what I know him or her to be like or what I think he or she ought to be like on the basis of my stereotyped preconceived notions?" While this is no easy task, it is necessary because there is no objective method to differentiate between stereotypes which are true and stereotypes which are false, and between natural stereotypes and those which are potentially socially harmful.

Exercise 11.2— How Are You Perceived?

The following exercise focuses upon your interaction with other individuals. It may help you think about how you behave when you initiate a relationship with another person or how you act in a group. The words below describe some of the ways people behave and perceive interactions. How are you described when interacting with others? Check at least five words that best describe you. Do people perceive you the same way you see yourself? Do people misjudge you? Discuss why you believe there might be a difference,

—Caring	—Unapproachable
—Cold	—Evasive
—Warm	—Judgmental
—Sensitive	—Resisting
—Agreeable	—Withdrawing
—Analytical	—Cooperative
—Sharing	—Helpful
—Initiative	—Directive
—Willing to concede	—Take-charge

NEED TO CHANGE ANY PERCEPTIONS? WILL THESE PERCEPTIONS HELP OR HINDER YOUR INTERPERSONAL RELATIONSHIPS WITH OTHER CULTURES?

A minority group refers to a category of individuals who are somehow distinguished from that of the general population by some socially-visible factors such as race, sex, or religion. Although this term has traditionally referred to a numerical minority comprising only a small fraction of the population, in recent years the concept has acquired more and more political connotations. Nowadays, a minority group may contain more or fewer people than the majority group. It is important to note that numbers of people are not as important as the power and status they possess.

American women, a numerical majority, have status problems that are very similar to blacks, a numerical minority. In both cases, the term minority group is appropriate.

The terms and concepts in this chapter are presented to enlighten your understanding of dealing with diversity. This campus offers a setting that is rich in diversity and can act as your human relations laboratory as you strive to reach your educational goals.

Minority groups

Name _____   Date _____

Exercise 11.3

Learning in a Multicultural Setting

Part I

Directions: Find out the different ethnicities and nationalities in your class. After you have consulted with your classmates and instructor, list them below.

_____

_____

_____

_____

_____

_____

_____

_____

Part II

Name and/or state five things you know about each ethnicity/nationality you named above. Include at least one famous person for each group. Write in the space below.

Ethnicity/Nationality   Things I know to be true

_____   _____

_____   _____

_____   _____

_____   _____

_____   _____

_____   _____

_____   _____

Part III

The information you thought was true was checked out with each group by

Part IV

Discuss what stereotypes, perceptions or assumptions emerged while doing the exercise. If you know that some of your information was inaccurate, take the time to clear it up. Respond to the following questions.

1. Name the stereotypes that were evident. Briefly explain your answer(s).

2. Do stereotypes help or hinder inter-ethnic communications.? Briefly explain your answer.

Part V: The Language Issue—Group discussion

Many people feel that people speaking another language in their presence, knowing they don't understand, are prejudiced and consider the behavior offensive. Read all questions before you begin your discussions.

1. What are your thoughts on this issue?

2. Do you believe others migrating to a new country should learn and speak the language of the new country?

3. What would you do If someone was speaking in your presence and you could not understand?

4. Picture yourself taking a required class you need in order to graduate at the end of the term and your professor has an accent that is difficult to understand. Besides dropping the class, how would you handle it? It is already too late in the term to get a refund for the course. If you do not pass this course, you will not be able to graduate.

Name _____   Date _____

Exercise 11.4

COPING WITH DIVERSITY

Now that you are aware of where you live and the demographers' predictions about racial/ethnic make-up of the United States population, do a personal inventory by answering the following questions. It should help you in determining whether you are ready for the challenges ahead.

Personal Inventory

1. How has your race/ethnicity affected your life to date?

    _____

2. How do you think your race/ethnicity will affect your choice of careers?

    _____

3. Share one thing that you are proud about in terms of your ethnicity. Why?

    _____

4. Do you readily share your ethnicity with others? Why? Why not?

    _____

5. Do you think that your color is going to affect your choice of careers? Briefly explain your answer.

    _____

6. Will your chances for upward mobility be affected by your sex, language or racial/ethnic makeup? Why? Why not?

    _____

7. Do you believe that one race is superior to another? Yes? No? Why?

    _____

8. Where do you plan to live? _____ What do you need to know and do to comfortably fit in?_____

9. How many languages will you need to speak in the future? Why?

    _____

10. Do you feel uncomfortable talking about racial issues? _____

11. Explain how you came to feel how you do about diversity issues.

    _____

Now analyze your answers. Are you ready for the future?

Human Relation Tips

These tips have been modified from Walker and Brokaw's "Thirteen Commandments of Human Relations" from the text, *Becoming Aware*. Since we are sure to encounter some conflicts along the way, perhaps these tips will aid us in our interactions with others.

Tip 1:   Greet people.

Speak to people. There is nothing as nice as a cheerful word of greeting.

Tip 2:   Smile.

When all else fails, smile. Smile at people. It takes 72 muscles to frown, only 14 to smile.

Tip 3:   Address a person by his/her name.

Call people by name. The sweetest music to anyone's ears is the sound of his/her own name.

Tip 4:   Be a friend.

Be friendly and helpful. If you would have friends, be friendly.

Tip 5:   Be cordial.

Speak and act as if everything you do were a genuine pleasure.

Tip 6:   Be real.

Be genuinely interested in people. You can like something in everybody if you try.

Tip 7:   Give praise.

Be generous with praise—cautious with criticism.

Tip 8:   When you don't know, ask.

Don't make assumptions; ask. When possible go directly to the source. This will help in eliminating stereotyping.

Tip 9:   Be aware that feelings do count.

Be considerate of the feelings of others. It will be appreciated.

Tip 10: There is always more than one point of view.

Be thoughtful of the opinion of others. There are three sides to controversy—yours—the other fellow's—and the right one.

Tip 11: Body language counts.

Be aware of how your body language affects others. Very often body language cancels out verbal language.

Tip 12: Give to others.

Be alert to give service. Lending a helping hand is what we should do for people since it tends to be so important in life.

Tip 13: Use tact.

When you speak, be careful how you say it and when you say it. There is more than one way to get your message across.

 Journal Questions/Activities

1. Find out the latest demographic changes for the United States by going on line to website `http://www.census.gov/population`.

2. Personally, how do you believe diversity is going to affect you? Explain your answer.

3. Share and explain your guiding principles to the whole issue of diversity.

Summary

In this chapter we have covered essential information you need to know if you plan to continue to live and work in this diverse society. We have included the current U. S. demographics and what it is projected to be midway in this century. Also included in this chapter are term definitions usually associated with discussions on diversity. The terms include: diversity, ethnicity, race, racism, prejudice, discrimination, scapegoating, anti-semitism, stereotyping, culture, minority groups and American values. We believe that a close analysis of these terms will help to enhance not only your college experience but help you to communicate effectively, live in harmony and work with individuals of different racial, ethnic and physiological makeups. There are bound to be problems and we are unable to prepare you to work with each of them. However, as a closing to this chapter, we presented some simple rules that will foster good inter-ethnic relations. These rules have been modified from Walker and Brokaw's "Thirteen Commandments of Human Relations" from the text *Becoming Aware*.

Summary Exercise 11.5

DIRECTIONS: Fill in the blanks with the appropriate word from the choices provided.

1. _____ couples the false assumption that race determines psychological and cultural traits with the belief that one race is superior to another.

   a. Discrimination   b. Racism   c. Ethnic groups   d. Prejudice

2. _____ is differential treatment based on unfair categorization.

   a. Scapegoating   b. Discrimination   c. Racism   d. Prejudice

3. _____ is a group whose members identify with one another because they share a common culture such as race or religion.

   a. Minority group   b. Ethnic group   c. Culture   d. Racism

4. _____ refers to the deliberate policy of blaming an individual or group when the fault actually lies elsewhere.

   a. Discrimination   b. Stereotyping   c. Scapegoating   d. Racism

5. _____ is prejudice or discrimination against Jews based on negative perceptions of their religious beliefs and/or on negative group stereotypes.

   a. Racism   b. Discrimination   c. Anti-semitism   d. Prejudice

6. _____ is the process of lumping perceptions into broad categories and processing them on the basis of what the categories are like rather than the unique characteristics of each individual.

   a. Scapegoating   b. Stereotyping   c. Discrimination   d. Racism

7. _____ is a set of rigid and unfavorable attitudes which are formed in disregard of facts toward a particular group or groups.

   a. Racism   b. Discrimination   c. Prejudice   d. Stereotyping

8. _____ is the way of life of a group of people which includes all the behavioral traits and material artifacts that are used by the people of a society.

   a. Minority group   b. Ethnic group   c. Culture   d. Diversity

9. _____ are a category of individuals who are somehow distinguished from that of the general population by some socially visible factors such as race, sex, or religion.

   a. Ethnic group   b. Women   c. Minority groups d. Culture

10. _____ refers to differences in regard to male and female, black and white, Hispanic, African, Asian, Native American, young, old, disabled, socioeconomic status, and/or religious preference.

   a. Culture   b. Discrimination   c. Diversity   d. Ethnic group

References

ADL. "A Campus of Difference" Workshop, Miami-Dade Community College. Miami, Fla.

Armas, Genaro C. "Hispanic growth viewed as opening for 2 top minority groups." *The Her* 2001) 26A.

Brislin, Richard W. and Pederson, Paul. *Cross-Cultural Orientation Programs*. New York: Press, Inc., 1976.

Department of Student Development, Miami-Dade Community College/North Campu Connection: An Interpersonal/Interethnic Experience." Miami: Unpublished. 1979.

Gardner, John and Jewler, A. Jerome. *College Is Only The Beginning: A Student Guide t Education*, 2nd edition. California: Wadsworth Publishing Co., 1989.

Gillet-Karon, R., Roueche, S., and Roueche, J. E. "Underrepresentation and the Que Diversity." *Community, Technical, and Junior College Journal*. 1990-91, 61, 22-25.

http://factfinder.census.gov. 2004.

http://www.census.gov. 2004.

http://www.independentamericans.org.

International Association of Business Communicators, The, *Without Bias: A Guide Nondiscrimincatory Communication*, second edition. New York: Wiley Publishing, 1982

Kappner, Augusta. "Creating-Something to Celebrate: Planning for Diversity." *Co Technical, and Junior College Journal*. 1990-91, 61, 16-21.

Keyser, John and Keyser, Marilyn. "From Affirmative Action to Affirming Diversity." *Co Technical, and Junior College Journal*, vol. 61, No. 3, Dec./Jan. 1990/91.

Miami Herald, The. Nation Briefs. Washingon, D.C., Wednesday, December 31, 2003, p.

Pearlston, Carl. "Is America a Christian Nation?" *Connecticut Jewish Ledger* (2001). Website /catholiceducation.org/articles/politics/pq0040.html.

Starke, Mary C. *Survival Skills for College*. New Jersey: Prentice Hall, 1993.

Tolerance. On-line Dictionary. http://www.merriam-webster.com.

Walker, Velma and Brokaw, Lynn. *Becoming Aware: A Human Relations Handbook*, 4th edit Kendall-Hunt Publishing Co., 1981, p. 146.

Watts, Patti. "Bias Busting: Diversity Training in the Work Place." *Management Review*. [1987, 51-54

Chapter Twelve

Wellness

Mary Mahan, Ed.D
Sandy Schultz, Ph.D.

Exercise 12.1

Wellness Awareness Check

DIRECTIONS: Place a check mark in the appropriate "true" or "false" box and see how you rate yourself on "wellness" awareness.

	True	False	
1.	❐	❐	Regular exercise can reduce stress and increase quality of life.
2.	❐	❐	Performance in school, sports, and work can be affected by the kind of foods you eat.
3.	❐	❐	Nutritionally good foods cost more than nutritionally inadequate foods.
4.	❐	❐	A college student should eat two to four servings of fruit every day.
5.	❐	❐	Carbohydrates are the body's most important source of energy.
6.	❐	❐	The substance most abused by college students is cocaine.
7.	❐	❐	One alcoholic drink is equal to one 12-ounce beer, or one 5-ounce glass of wine, or one mixed drink with 1.5 ounces of liquor.
8.	❐	❐	Of all infectious communicable diseases, sexually-transmitted diseases rank highest with students of college age.
9.	❐	❐	Many communicable diseases can be prevented by immunization.
10.	❐	❐	A regular systematic program of exercise increases your total blood cholesterol which aids the body's cardiovascular system.

Introduction

Active participation in a Wellness Program can positively affect the length and quality of your life. This participation includes a combination of sound nutritional habits, stress management techniques, ideal body weight maintenance, avoidance of substance abuse, knowledge of disease prevention, and a regular systematic program of combined aerobic exercise, resistance/strength/training, and flexibility training. Adopting these positive lifestyle practices while in college will help make you responsible for your own health care. Developing intelligent and lifelong wellness practices is what this chapter is all about.

Positive lifestyle practices

Nutrition

College students have many demands placed upon them. Performance in school, sports, work, and leisure time activities can be affected by the foods you eat or don't eat. Only you can control the amounts and types of food that you consume. Just as a car won't run well with poor quality fuel, your body will eventually suffer if you deprive it of essential nutrients. In addition, poor eating habits may increase your risk of heart disease, certain types of cancers, diabetes, stroke and osteoporosis.

Nutritionally good food costs no more than nutritionally inadequate food and can actually cost less. Having a thorough knowledge of nutrition can enable you to choose foods which enhance your diet and give you the best food value for your money. Following the ten Dietary Guidelines for Good Health of the U.S. Department of Agriculture will help you to improve your nutrition and can help prevent dietary problems.

1. Aim for a healthy weight.

    Being overweight or obese can increase your risk of high blood pressure, high blood cholesterol, heart disease, stroke, certain types of cancers, and other health problems. Your should evaluate your weight as it relates to your body fat percentage, blood pressure, and blood cholesterol levels. It may be necessary to reduce your weight if any of these measurements is high. Consult your physician first for advice regarding a weight loss program.

2. Be physically active each day.

    Make physical activity part of your daily routine. Try to accumulate a minimum of 30 minutes of moderate physical activity each day. Increasing the amount of time that you are active each day or participating in vigorous activities can give you added health benefits.

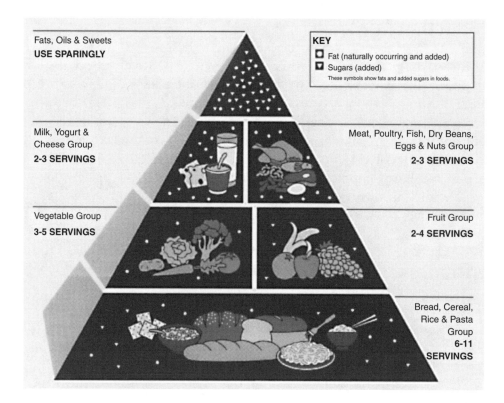

3. Use the Food Guide Pyramid.

    The Food Guide Pyramid can help you get a sufficient amount of nutrients and is compatible with all ethnic menus. Use the lower number of servings if you don't need as many calories and be aware of the amount of food that counts as a serving.

    People who choose not to eat animal products need to make sure that they get enough nutrients from the other food groups. Some people may have a greater need for certain nutrients. For example, adolescents and adults over 50 have a higher need for calcium, women of child bearing age need good sources of iron, and women who could become pregnant need extra folic acid.

4. Choose a variety of grains daily, especially whole grains.

    Grains are the foundation of a nutritious diet, providing you with vitamins, minerals, and carbohydrates (starch and dietary fiber). Grains are low in fat (unless fat is added during processing or preparation) and the high fiber content may help you feel full with fewer calories. Fiber-rich foods (whole grains, fruits and vegetables) also help promote regular bowel function. You should eat least six servings of grain products daily.

5.   Choose a variety of fruits and vegetables daily.

Certain fruits and vegetables are rich in Vitamin C and others may be good sources of carotenoids, including those which form Vitamin A. Many fruits and vegetables, especially dry beans and peas, are also high in fiber, naturally low in fat and calories, and help you feel full. Try to increase your intake of nutrient-rich dark, green leafy vegetables, deeply colored fruits and dried peas and beans.

6.   Keep foods safe to eat.

   • To reduce your risk of foodborne illness, wash your hands, utensils, and counters with hot, soapy water.

   • Separate raw, cooked, and ready-to-eat foods while shopping, preparing, or storing foods.

   • Cook foods to a safe temperature and refrigerate perishable foods promptly, following directions on the label. Serve hot foods hot and cold foods cold to reduce harmful bacteria growth.

   • When in doubt, throw it out.

7.   Choose a diet that is low in saturated fat and cholesterol and moderate in total fat.

Fats are an important part of our diet, supplying energy and essential fatty acids and helping absorb fat soluble vitamins A, D, E, K and carotenoids. Certain fats, especially saturated, increase your risk of heart disease by raising blood cholesterol levels. Unsaturated fats (vegetable oils) are more healthful, as they do not increase blood cholesterol. However, all fats are high in calories.

Your total fat intake should not be more than 30 percent of your total daily calorie intake. To reduce intake of saturated fat and cholesterol, limit use of solid fats such as butter, hard margarine, lard, and partially hydrogenated vegetable shortening. Substitute with healthier vegetable oils such as olive, canola, and peanut oils. Fat free and low fat dairy products, dry beans and peas, fish, lean meat, and skinless poultry are good choices for low fat, low cholesterol foods.

8. Choose beverages and foods to moderate your intake of sugars.

Carbohydrates are the body's primary source of energy and are found in milk, fruits, vegetables, breads, cereals, and grains. Simple sugars such as those found in fruits, candy, jellies, and donuts are quickly broken down and absorbed by the body. Complex carbohydrates (starches) break down into sugars more slowly, providing a more stable form of energy, plus they are rich in vitamins, minerals, and fiber.

Many foods, such as soft drinks, candy, cookies, cakes, and certain fruit drinks, are major sources of sugars that were added during processing or preparation. These added sugar foods provide calories but may have few vitamins and minerals, thus consuming excess calories from these foods may cause weight gain or prevent consumption of more nutritious foods.

9. Choose and prepare foods with less salt.

Consuming less salt is recommended for the healthy, normal person. Benefits of lowering your salt intake include reduced chances of developing high blood pressure and a decreased risk of calcium loss from the bones. Include fruits and vegetables (naturally low in salt unless it was added during processing) in your diet and read the Nutrition Facts Label on the package to identify foods lower in sodium.

10. If you drink alcoholic beverages, do so in moderation.

Alcoholic beverages supply calories but few nutrients. Consuming large amounts of these "empty calories" from alcohol may result in malnutrition in heavy drinkers, if they substitute alcohol for more nutritious foods.

Drinking in moderation is defined as one drink per day for women and two drinks per day for men, based on differences in weight and metabolism. Twelve ounces of regular beer, 5 ounces of wine, or 1.5 ounces of 80 proof distilled spirits count as one drink. Drinking more than one or two drinks per day raises your risk for motor vehicle crashes, other injuries, high blood pressure, stroke, violence, suicide, and certain types of cancer. Women should never drink during pregnancy as alcohol increases the risk of birth defects. Other problems related to excess alcohol consumption include cirrhosis of the liver, damage to the brain and heart, and social and psychological problems.

Stress

Stress—without it, life would be boring! Stress is the body's response to demands made upon it by physical or psychological stimuli. Positive stress, called eustress, results in better health and improved performance. Winning the lottery, receiving a promotion at work, or becoming engaged are examples of eustress. Distress, which occurs when responding to negative stressors such as a failing grade, loss of a job, death of a loved one, or a divorce can be accompanied by deterioration in health and poor performance.

Your body responds to eustress and distress in a similar manner. You may experience some of the following temporary effects of stress on your body:

- increase in heart rate

- rapid breathing and/or shortness of breath

- constipation or diarrhea

- lower back pain

- tiredness

- headaches and/or dizziness

- sleep problems

- irritability and/or moodiness

- inability to concentrate

If you are unable to deal with stress for a prolonged period of time, serious physical and mental problems such as stomach ulcers, heart disease, severe headaches, hypertension, depression, weight problems, and drug and/or alcohol abuse may develop. Research indicates that people who are very stressed seem more likely to catch an infectious disease.

MANAGING STRESS

You must recognize that stress is a problem in your life before you can deal with it. Life is full of daily hassles such as misplacing your keys, getting stuck in traffic, waiting in lines, and experiencing other minor annoyances at school, work, and home. When you realize that these situations are not worth getting upset over, you will learn to put up with them and be proud of the fact that you have control over your emotions.

Sometimes stress is difficult or not possible to control. In those cases, you must learn to cope with the stress.

TECHNIQUES FOR MANAGING STRESS

- Participate in activities which are enjoyable for you. This is important to your well-being. Make time each day in your life for fun. Laughter is also a great stress reducer.

- Exercise can be a great stress-reducer. People who exercise regularly are able to handle stress better.

- Release emotions in a positive manner. Crying is a very healthy way to release emotions, as long as it is not excessive.

- Practice good time management techniques. Organize your time by keeping an appointment calendar indicating important dates, events, and assignments.

- Learn to say "no" in a tactful manner if too many demands are being placed on you.

- Set reasonable goals for yourself and write them down. Challenging but realistic goals will keep you motivated and help you stay on track to achieve those goals. Goals which are too difficult will only add to your frustration and stress. Don't be afraid to modify your goals if necessary.

- Eat a healthy, balanced diet to give you energy to help cope with stress. Cut down or avoid caffeine.

- Talk with someone such as a counselor, teacher, family member or friend whom you trust. Sharing your problems and concerns with another person helps, and he/she may offer another view of the problem. If you feel very distressed, overwhelmed, or depressed, possible sources of help are your doctor, school psychologist or counselor, and local health agencies.

- Practice techniques such as tai chi, deep breathing, meditation, yoga, massage, imagery, or progressive muscle relaxation.

Remember that the way in which you react to the stressor, not the stressor itself, is the cause of many stress-related illnesses. Take the time to learn which stress-reduction techniques work best for you.

Infectious Diseases

Ways to decrease chances of contracting infectious diseases

College students should become informed about the many different kinds of infectious diseases that exist in the world. Attending college for the first time translates into more independence for the student and, consequently, less observation and monitoring by parental authority than in the past. Additionally, frequenting a new environment populated by a large number of out-of-state and international travelers can increase the chances of contracting infection. Practicing a healthy lifestyle, washing hands frequently. and obtaining immunizations are ways to raise the body's immune system and prevent infectious disease.

Infectious diseases are caused by microorganisms called germs. Fortunately, very few microorganisms are disease-producing in humans. Besides, germs must find a way to be transmitted and find entry into a human being to be pathogenic (disease-producing). Disease-producing germs come in various sizes from microscopic sizes to almost visible forms. Virus are the smallest germs while Richettsia are barely detectable under a microscope. Bacteria germs come in three micro shapes: rod, spherical, and spiral. Fungi germs include molds and yeasts, and other plantlike microorganisms. Protozoa are single-celled parasite germs while worms are larger, multi-celled animals.

Virus

Infectious diseases can be transmitted by several common ways: respiratory discharge, discharges from the intestinal tract, contaminated water or soil, contaminated food or milk, association with animals, insect bite, intimate contact, and sexual activity. Some infectious diseases can be prevented by artificial immunization (inoculations), but many more cannot be controlled by this means. Being alert to the danger signals of the onset of various diseases and seeking early medical treatment are prudent practices to follow. Avoiding the causes of such diseases is also an important practice.

Transmission of infectious diseases

Sanitary environmental conditions and the development of antibiotic drugs and penicillin have improved the treatment of individuals who have symptoms or have been diagnosed as having an infectious disease.

SEXUALLY-TRANSMITTED DISEASES

Of all the infectious diseases, sexually transmitted diseases (STDs) rank highest with students of college age. STDs are caused by sexual contact or intercourse as one person infects the other(s). They occur again and again because the body can't build immunity against them. The five most common STDs are gonorrhea, syphilis, Acquired Immune Deficiency Syndrome (AIDS), chlamydia, and genital herpes. Other STDs are genital warts, hepatitis B, and trichomoniasis.

Gonorrhea is caused by a bacterial infection. Painful urination and pus discharge from the penis in males, and possible minor urinary discomfort and/or vaginal discharge in females are the typical symptoms. Treatment of choice for gonorrhea is penicillin. Its symptoms show up between two and twenty-one days after having sex, however, most women and many men have no symptoms. Gonorrhea can damage reproductive organs and cause heart trouble, skin disease, and blindness.

Gonorrhea

Syphilis, one of the most dangerous STDs, is a four-stage disease caused by spirochetes. In the initial stage, a lesion appears on the genitals while a genital rash is common in the second stage. The third stage is one of latency with no symptoms; however, syphilis remains highly contagious. The final stage is one of tissue destruction and possibly death. Treatment includes maintaining high levels of penicillin, erythromycin or tetracycline in the blood stream for a specified period of time until all spirochetes are dead. Syphilis can cause heart disease, brain damage, blindness, and death.

Syphilis

AIDS   Acquired Immune Deficiency Syndrome (AIDS) caused by the Human Immunodeficiency Virus (HIV) can be transmitted during sexual contact, via body fluids, or by sharing needles. The HIV attacks white blood cells in the human blood, weakening the immune system and damaging one's ability to fight off other invading diseases. Currently, there is no vaccine to prevent AIDS; nor is there any proven AIDS cure, but there are effective treatments available to delay the onset of AIDS. In the early stages of AIDS, there are no physical symptoms or signs that indicate a person has been infected. With a weakened immune system, individuals are subject to infection by various other diseases or to damage to the nervous system and brain by the AIDS virus itself. Death will eventually occur. Mothers can give AIDS to their babies in the womb, during birth or while breastfeeding.

Chlamydia   Chlamydia is a bacteria-like microbe with some characteristics of a virus. It can be mistaken for gonorrhea. Chlamydia must be treated with tetracycline rather than penicillin. Its symptoms show up between seven and twenty-one days after having sex, however, most women and many men have no symptoms. Clamydia can lead to more serious infection, causing damage to reproductive organs.

Genital herpes   Genital herpes is caused by a virus which promotes sores near the infected genitals (type 1) or on the labial area (type 2). Herpes simplex virus type 1 or type 2 can be treated without drugs. However, herpes cannot be cured or completely eradicated unlike most other STDs. Symtoms show up between two and thirty days after having sex. Most women and many men have no symptoms. Herpes is spread during sexual intercourse, or oral and anal sex with someone who has herpes.

Hepatitis B   Hepatitis B is an infection of the liver spread the same way as HIV. There is no medicine that can cure hepatitis but interferon alfa-2B helps some individuals. Sound nutrition and adequate rest are important for anyone with hepatitis.

Genital warts   Genital warts are caused by genital human papilloma viruses. Pink or reddish warts with cauliflower-like tops appear on the genitals or in the mouth. Surface medications and laser surgery have proven effective.

Trichomoniases   Trichomoniases is a common form of vaginitis in women causing a strong odorous yellowish vaginal discharge that is itchy. Men have few or no symptoms. The drug, metronidazale, taken by both partners has proven effective in treatment.

Safer Sex   Control of certain human behaviors is essential to the prevention of the spread of STDs and AIDS. Abstinence from sexual intercourse and/or intravenous drug use are preferred behaviors. Faithful monogamous relationships and "safer sex" practices—use of latex condoms combined with the spermicidal chemical Nonoxynol-9 from the beginning to the end of sexual intercourse—are desirable behaviors. Oral sex should not be performed when either partner is considered at high risk. College students must understand and put into practice sexual behaviors which are prudent, conscientious, and healthy. Once symptoms of STDs occur, early testing, counseling, and treatment are recommended.

Substance Abuse

Admission to and attendance in college is a new and exciting experience for students. New friendships and acquaintances are formed. Time is spent on campus, in the classroom, and at organized activities where students are faced with the many influences of campus life which can result in either positive or negative impacts. The use of drugs, steroids, alcohol, or tobacco is a dangerous practice for college students. Research indicates that the influence of friends is the most-cited reason for experimentation in substance use. **Influence of friends**

The most important factor regarding substances is not the question of legality, but rather the effect a substance or combination of substances has on the mind, body, and life of a college student. The key to abstaining from the use of substances is to develop sufficient will power rather than to rely on imposed external forces. Will power can be developed just as one develops sound study habits. However, a genetic predisposition to substance abuse may overwhelm an individual's will power. In these instances, external safeguards may be helpful in substance abuse control for certain individuals. **Will power**

Knowledge, emotional maturity, and will power are qualities which will help students avoid harmful substances whether the substances are considered to be prescription, illegal, addictive, synthetic, natural, dietary, or social. **"Just say no!"**

College students should keep in mind that alcohol is the most abused of all substances. Its low cost, legal status and easy availability make it popular among college students and, therefore, subject to abuse. Vandalism, date rape, academic problems, dropouts, injuries, missing class, and death have been attributed to alcohol abuse among college students. Alcohol use signifies the emergence from youth to adulthood, enhances social gatherings and helps to cope with stress. Unfortunately abuse can lead to binge drinking as well as unplanned and unprotected sex. **Alcohol**

Use alternatives to drinking alcohol in social situations and while being alone. Exercise can substitute for alcohol use to control stress, serve as a coping mechanism, and can be a great social activity. Groups like Alcoholics Anonymous and Al-Anon can be of support to a problem drinker.

Steroids are synthetic derivatives of the male hormone testosterone. Students take steroids to help produce large muscles. Steroid-takers are subject to aggressive behavior, high blood pressure, cardiovascular disease, cancer and a long litany of other side effects. The irony is that the intended larger muscle tissue that is produced is highly susceptible to injury by the steroid taker. **Steroids**

Remember—your heart is a muscle too! Steroids are taken in a series that are ingested and/or injected. Steroid use is illegal, is a felony and can cause damage to reproductive organs as well as the immune system.

Tobacco   Tobacco products, whether smoked, chewed, or dipped, are worldwide health hazards accounting for tens of thousands of deaths annually. Lungs, mouth, pharynx, esophagus, heart and circulatory system, skin, immune system and other systems and organs are subject to free radical damage caused by the use of tobacco products.

Cigars are not safe alternatives to cigarettes. One or more smoked cigars each day have the same risks of causing cancer and heart disease as cigarettes. A major health concern is exposure to second-hand smoke in the homes of infants and children, who are actually at a higher risk than non-smokers for tobacco-associated diseases.

Seek help to give up smoking by participating in a smoking cessation clinic at your wellness center.

Drugs   Drugs enter the body by injection, inhalation, ingestion, and absorption through the skin or mucous membranes. While the use of cocaine, crack, LSD, and marijuana has decreased among college students you now must become "safe party goers." Knowledgeable of the dangers of recreational/club drugs such as Ecstasy and Ritalin as well as the date rape drugs GHB and Rohypnol is necessary. Sexual contact under the influence of alcohol and these drugs increases the risk of HIV and STD infections.

Ecstasy   Ecstasy (XTC) (MDMA) produces a sense of well being, energy and sexual stimulation and is popular at rave parties. It can cause damage to the liver and perhaps the brain, drains water from the spine, and interferes with normal nerve function. The real danger occurs in its chemistry when mixed with alcohol, dairy products, or chocolate.

Ritalin   Ritalin (MPH) (Vitamin R, R—Ball) is a mild stimulant commonly prescribed for young children to treat attention deficit/hyperactivity disorder (ADHD). Used recreationally, the drug can give a sense of euphoria, particularly when it enters the bloodstream quickly. Abusers will inject or inhale the drug, which is ranked among the top ten controlled pharmaceuticals most frequently reported stolen.

Rohypnol   Rohypnol has been a concern as the "date rape" drug. College students can unknowingly be given the drug mixed in a drink and become incapacitated to resist sexual activity or can die if the drug is mixed with alcohol and/or other depressants. Rohypnol produces sedative hypnotic effects as well as physical and psychological dependence.

GHB   GHB (Liquid ecstasy, Scoop) is abused as a sedative body-building drug, and is associated with sexual assault. Coma and seizures can occur with GHB abuse. Combined with alcohol, nausea and difficult breathing may occur. Since the drug causes amnesia you don't remember what happened during the time you were on the drug.

Unless a drink is offered to you in a sealed container and opened personally, it should not be consumed at a club or private party. Keep your drink with you at all times.

Exercise—Nature's Tranquilizer

Aerobic Exercise

Aerobic exercise is continuous and rhythmic exercise performed for twenty to sixty minutes, three to five times a week. It takes up, transports, and delivers oxygen to the muscle.

Aerobics

Healthy college students should participate in aerobic activities (walking, running, cycling, in-line skating, swimming etc.) or work out on specific aerobic designed machines (treadmills, bikes, steppers, rowing machines, versa climbers etc.) three to five times a week. This systematic program of exercise should elevate the heart to a training zone intensity for 20-60 minutes. For weight loss aerobic exercise should be performed five times per week.

Exercising with a group or partner can help keep you motivated and can be fun. You should always be able to catch your breath, speak, or sing comfortably. Discomfort may be normal. Pain is not. Don't forget to warmup before and stretch and cool down for five to ten minutes following activity.

Exercise 12.2

Calculating Your Training Zone

1. Subtract your age from 220 = MHR

2. Multiply your MHR x .60 = lower limit

3. Multiply your MHR x .85 = higher limit

   Your training zone is _____ to_____

Exercising at the lower end of your zone for a longer time is better than at the high end for a shorter time if you want to burn more calories from fat.

Your physical education instructor, wellness director, or certified personal trainer can assist you in monitoring your pulse and/or heart rate before, during, and after exercise. He/she can also demonstrate safe body alignment during performance and assist you in understanding the scientific principles involved in selecting and performing the appropriate exercise program for your particular body shape, size, and needs.

Resistance/ strength training

Resistance/strength training programs are designed to increase the strength, size, and endurance of fibers that make up muscles. Utilizing all the major muscle groups of the body, resistance weight trainees perform eight to 12 repetitions per set on strength training machines with proper form and with a full range of motion during workouts two or three times per week. A variety of different training methods can be developed. Combining aerobic and resistance weight training with a daily regimen of stretching movements is recommended for a total training program for college students. Aerobic exercise, resistance weight-training, and flexibility training are for almost everyone. Students with high blood pressure should participate in weight training only with the consent of their physician. Beginning or accomplished exercise enthusiasts can improve their flexibility, balance, strength, endurance, and respiratory/circulatory systems as well as maintain a lean body during resistance/strength training.

Staying strong and flexible is the key to staying active in later years. Basic everyday movements such as getting up from a chair require muscular strength. Strength is a benefit at any age and any level of fitness. The development of muscle mass is important in losing and maintaining weight. Lean body tissue (muscle) expends calories at a faster rate than fat tissue. Certain chronic problems such as lower back pain are often related to poor abdominal muscle strength and inflexible posture muscles (hamstrings). Getting stronger and maintaining flexibility can improve your performance in sport, exercise, and dance activities.

Other specific benefits of a regular systematic program of exercise include

- Increased self-confidence and self-esteem

- More energy

- Improved circulation and lower blood pressure

- Reduced tension and assistance in stress management

- Lowered resting heart rate (aerobic training)

- Decreased total blood cholesterol (aerobic training)

- Maintenance of proper weight.

The Importance of Sleep

You will probably spend about one third of your life sleeping. During sleep, the regeneration of body cells accelerates; thus, young people who are growing and older people who need more time to recuperate often require more sleep than others. Most college students find that six to eight hours of sleep per night is sufficient, but some require more or less.

Insomnia can be caused by many factors:

- stress brought about by physical, social, psychological or economic problems;

- excessive fatigue;

- excitement or anticipation of a trip or event;

- intake of caffeine, nicotine or other stimulants late in the day;

- eating or drinking excessively;

- vigorous exercise just before bedtime. Exercise early enough to allow several hours for your metabolism to slow down.

If you experience occasional problems of falling asleep, try any of the following:

- sleep in a dark, quiet room;

- go to bed at approximately the same time each night and get up at the same time each morning;

- keep the room temperature comfortable;

- sleep in a comfortable bed;

- drink warm milk or eat a light snack just before bedtime;

- take a warm bath a couple of hours before bedtime;

- listen to relaxing music or read a book;

- perform relaxation exercises.

? **Journal Questions/Activities**

1.   How can the Food Guide Pyramid help you improve your food choices?

2.   What methods can you use to improve the ways in which you deal with stress?

3.   Describe a realistic exercise programw that will help you achieve your fitness goals.

4.   In addition to changes in your nutrition, stress management, and exercise program, how else could you improve your wellness?

5.   Which topic in this chapter was the most meaningful to you? Why?

Summary

It is up to you to establish a lifestyle that will enable you to reach your potential. How you live now will have a great influence on the quality and length of your life. Many premature deaths occur in young individuals due to cardiovascular disease. These diseases of choice are related to our lifestyles and habits associated with lack of exercise, poor diet, stress and smoking. Would you like to know about how long you are going to live? Can you change some things about your lifestyle that might give you a better quality of life and possible a longer life? To find out, play the Longevity Game: http://www.nmfn.com/. Click into the Learning Center and find the longevity game.

Take advantage of opportunities to exercise daily. Join a fitness center or gym, take exercise classes such as aerobics, yoga, or swimming, learn a new sport. Make a commitment to yourself to improve your diet and reach your desired weight. Optimal wellness is a goal worth pursuing so that you may always enjoy a happy, healthy, and productive life.

Name _____   Date _____

Summary Exercise 12.3

DIRECTIONS: Answer the following questions.

1. How many servings of vegetables should you each each day?

2. Why is it important to reduce the saturated fat and cholesterol in your diet?

3. List three ways to reduce the amount of saturated fat in your diet.

    a.

    b.

    c.

4. List three stressors in your life and ways that you can manage them.

    a.

    b.

    c.

5. List two ways that infectious diseases can be prevented.

    a.

    b.

6. Name the substance that is most often abused by college students.

7. How can STDs be transmitted?

8. Why is it important to maintain flexibility?

9. Define aerobic exercise and list some of its benefits.

10. How can you reduce your risk of foodborne illnesses?

11. What is wellness?

References

American College of Sports Medicine 2000 ACSM's *Guidelines for Exercise Testing and Prescription*, 6th edition. Baltimore, Md.: Lippincott Williams & Wilkins.

American Dietetic Association 1999. *The Essential Guide to Nutrition and the Foods We Eat: Everything You Need to Know about the Foods You Eat.* New York: Harper Collins.

Fahey, T.D. 2000. *Basic Weight Training for Men and Women*, 4th edition. Mountain View, Ca.: Mayfield.

Florida Department of Health. What Everyone Should Know about STDs, Channing L. Bette Co., South Deerfield, Ma. 1999.

Florida Health and Rehabilitative Services. STD Control Program. Tallahassee, Florida. 2000.

Greenberg, J.S. 1999 *Comprehensive Stress Management*, 6th edition. Dubuque, Ia. Brown and Benchmark.

U.S. Department of Health and Human Services. *Surgeon General's Report of Acquired Immune Deficiency Syndrome.* Washington, D.C.: 2000.

U.S. Department of Agriculture and U.S. Department of Human Services. *Dietary Guidelines for Americans*, Washington, D.C.: 2000.

Websites

CARDIOVASCULAR RISK

American Heart Association	www.americanheart.org
National Heart, Lungs, and Blood Institute	www.nhlbi.nih.Qov
National Stroke Association	www.stroke.org

STRESS

American Psychological Association	www.apa.org
Center for Anxiety and Stress Treatment	www.stressrelease.com
National Institute of Mental Health	www.nimh.nih.gov
National Sleep Foundation	www.sleepfoundation.org

GENERAL FITNESS

Columbia University Health Service	www.goaskalice.columbia.edu
Shape Up America	www.shapeup.org
Fitness Link/Stretching Exercises	www.fitnesslink.com
Biomechanics Worldwide	www. per. ualberta.ca/biomechanics
Workout.wCom: Exercise Zone	www.workout.com/exercises
Physician and Sports Medicine	www.physsportsmed.com
American Alliance for Health, Physical Education, Recreation and Dance	www.aahperd.org
American Medical Association Personal Trainer	www.ama-assn.org/insight/gen-hlth/fitness

NUTRITION, WEIGHT CONTROL

U.S. Consumer Gateway: Health, Dieting,
    and Weight Control   www.consumer.gov/health.htm
Ask the Dietitian/Overweight   www.dietitian.com/overweig. html
American Dietetic Association   www.eatright.org
FDA Center for Food Safety and Applied
    Nutrition   www.vm.cfsan.fda. gov
USDA Center for Nutrition   www.usda.gov/cnpp
Vegetarian Resource Group   www.vrg.org

INFECTIOUS DISEASES

John's Hopkins Infection Diseases   www.hopkins-id.edu/id-nautop.htmi

DRUGS

National Institute
NIDA.   nih.gov/ of drug use-INFO FAX/hs
                                                    INFOFAX/HS

THE DAILY APPLE

the daily apple   Corn/target/CS Article/CS/html

HIV

McKinley Health Center
HIV Infection, Testing and Aids   McKinley.ufcu.edu/health
                                                    info/sexual/stds/hiv-qaa.html.

Appendix A

Student Information

Student No. _____ Social Security No. _____

Last Name _____ First Name _____ Date _____

Local Address

City _____ State _____ Zip Code

Phone No. _____ Sex _____ Race _____ Ethnicity

Marital Status _____ Ages of Children _____

Class Hours per Week _____ _____ Sem Crs.

Other colleges enrolled _____ Total Crs.

Employed _____ Hours per Week

Place of Employment _____ Job Title _____

Work Experience

Career Goals (circle one)

a. AA   b. AS   c. Non-degree

d. Armed Forces   e. Law Enforcement

f. Other _____

Barrier to Career Goals (circle appropriate responses)

a. Financial Aid   b. Work

c. Domestic (childcare, care of relative, relationship problem)

d. Health   e. Psychological   f. Low Motivation

g. Time Management   h. Low Academic Skills

i. Other _____

Hobbies (circle appropriate responses)

a. Sports   b. Reading   c. Music

d. TV/Cinema   e. Video game   f. Talking on phone

g. Sleeping   h. Shopping   i. Cruising

j. Other

Semester Schedule

Course Abbrev. No.	Location	Sequence No.	Days	Time	Instructor
1. _____	_____	_____	____	_____	_____
2. _____	_____	_____	____	_____	_____
3. _____	_____	_____	____	_____	_____
4. _____	_____	_____	____	_____	_____
6. _____	_____	_____	____	_____	_____
7. _____	_____	_____	____	_____	_____

Appendix B

Course Requirements

Course Number & Prefix; Title _____

Name of professor teaching the course _____

Course assignments & test dates

_____   _____

_____   _____

_____   _____

_____   _____

_____   _____

Write out your goal and grade expectation for this course:

Grade _____

Date(s) you plan to visit your professor

Appendix C

Student Development—Two-Year Academic Plan

STUDENT NAME _____ MAJOR _____ DEGREE _____

PROJECTED GRADUATION DATE _____

YEAR 1 AT M-DCC

1ST SEMESTER _____ DATE _____ | 2ND SEMESTER _____ DATE _____

(   )CREDITS TOWARD DEGREE | (   )CREDITS TOWARD DEGREE

3RD SEMESTER _____ DATE _____ | 4TH SEMESTER _____ DATE _____

(   )CREDITS TOWARD DEGREE | (   )CREDITS TOWARD DEGREE

YEAR 2

1ST SEMESTER _____ DATE _____ | 2ND SEMESTER _____ DATE _____

(   )CREDITS TOWARD DEGREE | (   )CREDITS TOWARD DEGREE

3RD SEMESTER _____ DATE _____ | 4TH SEMESTER _____ DATE _____

(   )CREDITS TOWARD DEGREE | (   )CREDITS TOWARD DEGREE

NOTE: PLEASE INCLUDE PREREQUISITES AND COREQUISITES WHEN LISTING COURSES.

CODING: P—PREREQUISITE   E—ELECTIVE   C—COREQUISITE

Index